BY INDUSTRY: WEBSITE MENU COLLECTION

BY INDUSTRY : WEBSITE MENU COLLECTION
業種別 ウェブメニュー コレクション

P・I・E BOOKS
4-14-6, Komagome, Toshima-ku, Tokyo 170-0003 JAPAN
TEL : +81-3-3940-8302
FAX : +81-3-3576-7361

Please note our change of contact details from 17th March 2003.
2-32-4, Minami-Otsuka, Toshima-ku, Tokyo 170-0005 JAPAN
TEL : +81-3-5395-4811
FAX : +81-3-5395-4812
e-mail : editor@piebooks.com
sales@piebooks.com
http://www.piebooks.com

ISBN4-89444-235-3 C3070
Printed in Japan

CONTENTS

はじめに

ウェブサイトを形成している要素で最も重要なものの1つにメニュー（目次）ページがあります。メニューページはユーザーに、このサイトにはどのようなページがあるのかを案内し、各ページに移動するための入り口的役割を担っています。ほとんどのサイトはこのページを持ち、TOPに近い階層に作られていたり、TOPそのものだったりするため、注目度が高く、デザイン性に優れたものが多く存在します。今回紹介しているメニューページには、大きく分けて2つのタイプがあります。

①目次項目が文字もしくはアイコン（絵・写真）で表示されており、それ自体がメニューボタンになっているもの。
②一見どれがメニューボタンかわからないが、どこかに隠れているもの。（メニュー表示が明らかでないもの。）

現在、ネット上では、圧倒的に①のパターンを多く見ます。
それは制作者にとって、メニューボタンやメニューバーを利用することで、サイトのメニューやリンク先を体系づけて構築しやすくなるという点からこのパターンで作ることが多くなったのでないかと考えられます。しかし、最も主な理由の一つとしては、ネット上にショップやメーカーなどの商業目的のサイトが多いということがあげられます。商品販売や、商品紹介、検索を目的としたサイトでは、いかにお客様側の操作性を良くするか、また、わかりやすいかを第一に考えるので、おのずと①のパターンを採用するところが多くなるのです。また、販売されているビギナー用ウェブサイト制作ソフトに、①の形態のデザインフォーマットがあらかじめ入っていたり、使用例などで紹介されることが多いというのも要因の一つです。アマチュア＝

ノンプロの場合は、簡単に自分のホームページを作りたいと考えるので、一般的なフォーマットを使います。このようにして毎日のように①のパターンのメニューページがネット上に増え続けているのです。本書においてもこのパターンのメニューページの方が多く紹介されています。初めてサイト作りをする人は、参考にしてみてください。

次に、忘れてはいけないのが、②のパターンです。一見わからないが、絵や写真のどこかの部分がメニューボタン（バー）になっているもの、更には、このページ自体がメニューページなのか何なのかすらわからないものも存在します。操作性や分かりやすさで考えれば、①のパターンと全く対極に位置するものですが、わかりにくい分、その中に隠されたものを見つけ出すといった宝探し的な楽しみ、思いがけない反応をした時の驚きは、くせになります。面白いサイトを探すことを生き甲斐にしているネットサーファー達が多いのもうなずけます。このパターンはデザイナーやアーティストなど個人のサイトでよく見られ、ウェブ制作の知識に富んだ人が、様々な技術を駆使して作っていることが多いようです。ありきたりのサイトに飽きた人は、掲載されているこのパターンの作品のアドレスにアクセスし、実際に体験してみてください。

ウェブサイトに興味があり、これから作ろうとしている人、現在公開しているサイトを更に改良したい人は、ぜひ全作品を通して見てください。きっとそこには、デザイン性に高い、優れたサイト作りのヒントが隠されているはずです。

ピエ・ブックス編集部

FOREWORD

One of the most important pages at any Web site is the menu page. It introduces the site's contents to users and is their gateway to the site. The menu page at most Web sites is either near the top page in the site's hierarchy or is itself the top page. Because menu pages are attention getters, there are many examples of design excellence to be found.

The menu pages contained in this volume have been broadly divided into two categories:

① Pages whose contents are displayed as text or icons (graphics, photos) that serve as menu buttons to navigate the site.

② Pages whose menu buttons are concealed or not immediately apparent to the eye.

Menu pages of the first type are by far the most prevalent on the Internet today. Menu buttons and bars offer Web site designers a simple way to systematize site menus and links. But the popularity of this type is principally due to the large number of e-commerce sites on the Internet. The top priority of product marketing, advertising, and search site designers is to create Web sites that are easy for customers to understand and navigate, so most adopt conventional methods.

The button-and-bar menu page is also the design format incorporated in Web site design software for novices and used for menu design tutorials and samples. The amateur, or non-professional, chooses the conventional format because it offers an easy way to design a homepage.

The Internet is bombarded daily by new sites with conventional menu pages. The category is well represented in this volume, and readers interested in designing their first Web site are encouraged to draw on these fine examples.

But let's not forget the second category of menu pages whose menu buttons and bars are not obvious at first glance—some are not even recognizable as menu pages at all! The type ② menu page is the polar opposite of its straightforward, easy-to-navigate counterpart. But the elusive nature of such pages is more than compensated for by the treasure-hunt experience and element of surprise they provide. It's no wonder there are so many Web surfers whose main satisfaction is hunting down yet another unique site.

The unorthodox menu page is often seen at the personal Web sites of designers and artists. Many have been created by Web designers expert in a wealth of technologies. Anyone bored with run-of-the-mill Web sites is encouraged to access the sites presented here and experience them firsthand.

This collection of superbly designed Web sites will be an inspiration to everyone interested in Web sites or planning to build or improve an existing a site. You will find some tips for creating excellent and stylish Web sites.

P•I•E BOOKS

EDITORIAL NOTES

CREDIT FORMAT

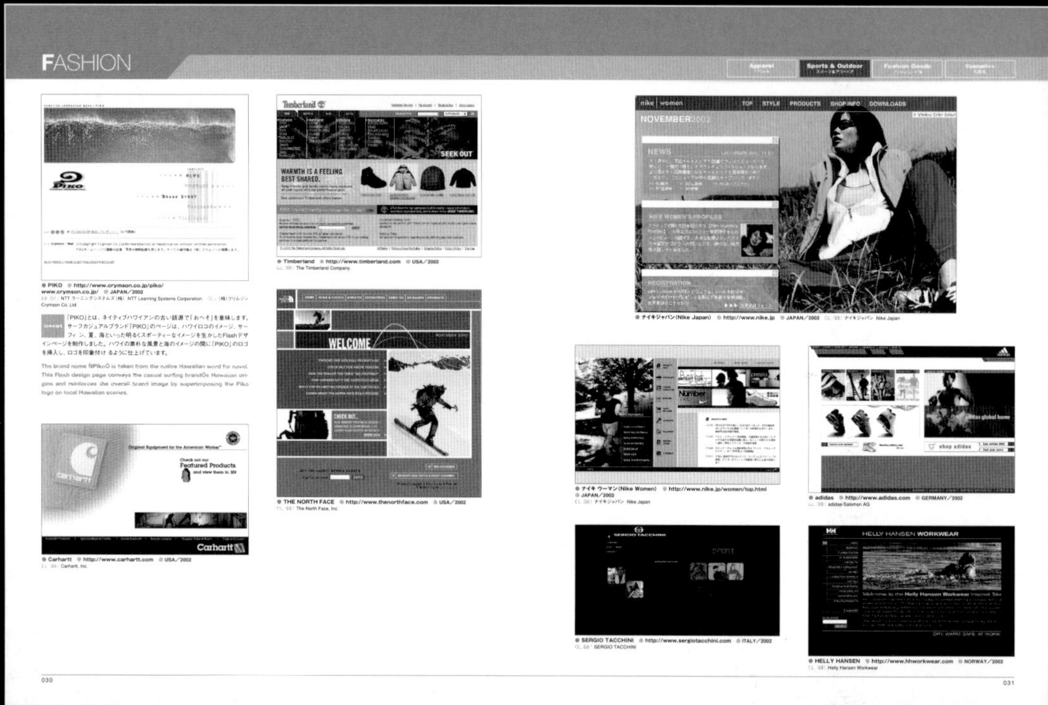

● **PIKO** ● **http://www.crymson.co.jp/piko/**
www.crymson.co.jp/ ● **JAPAN／2002**
SB, DF：NTT ラーニングシステムズ（株） NTT Learning Systems Corporation.　CL：（株）クリムゾン Crymson Co.,Ltd.

FASHION

Apparel
［アパレル］

Sports & Outdoor
［スポーツ＆アウトドア］

Fashion Goods
［ファッション小物］

Cosmetics
［化粧品］

● **AMERICAN EAGLE OUTFITTERS** ● **http://www.ae.com** ●

USA／2002

CL, SB：American Eagle Outfitters

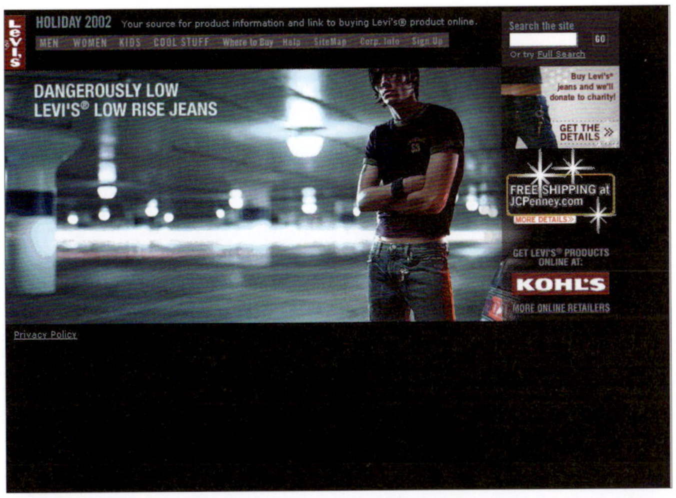

● Levi's ● http://www.levi.com ● USA／2002
CL, SB : Levi Strauss & Co.

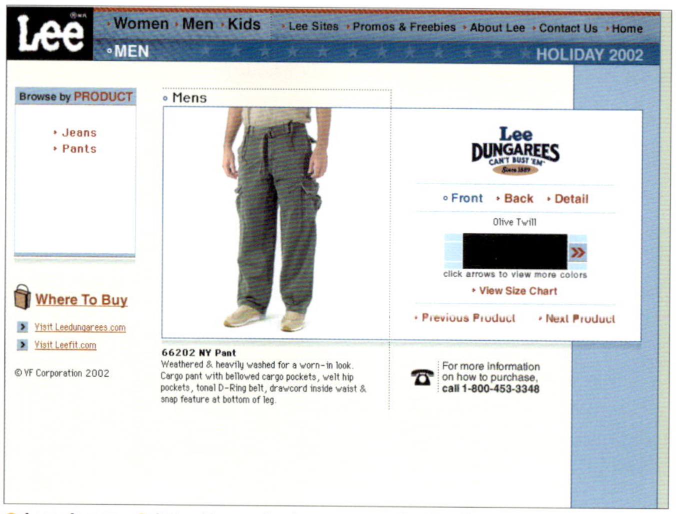

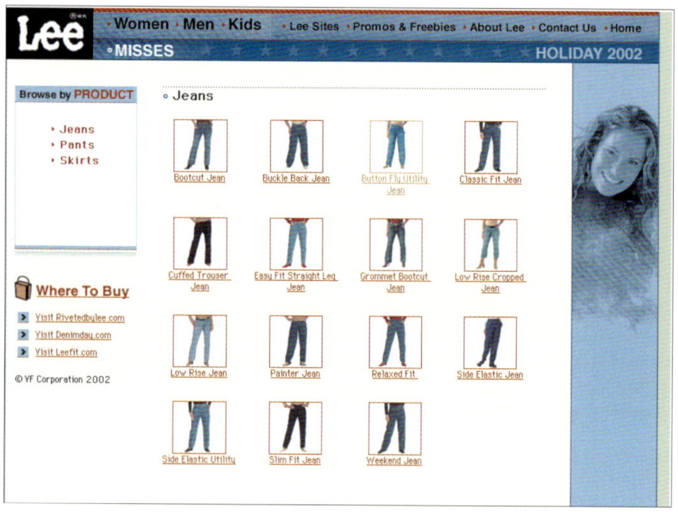

● Lee Jeans ● http://www.leejeans.com ● USA／2001

CL：Lee Jeans SB, DF：Lookandfeel new media

concept LeejeansのWebサイト。シンプルなマルチレイヤ式の使いやすいナビゲーションを使って、最初のページでカテゴリ(「Women（女性用）」、「Men（男性用）」、「Kids（子供用）」)、次のページで表示オプション(「Browse by Product（商品別表示）」、「Browse by Fit（サイズ別表示）」)を選択していくと、商品が表示される仕組みです。各ページの右側にはプロモーション用のスペースが設けられており、そのページのターゲットに応じた最新プロモーションの広告が表示されています。バックグラウンドのデニム生地をはじめ、Leeが描くライフスタイルのイメージがふんだんに組み込まれたデザインになっています。

We created an uncomplicated menu system for the Leejeans Web site that simplifies its multi-layered navigation. The site is divided into Women, Men, and Kids sections that users can browse by either Product or Fit. Each section also displays promotions for its clothing categories on the right side of the page. The design incorporates Lee's lifestyle images and a background of Lee Jeans denim fabric texture.

● GUESS ● http://www.guess.com ● USA／2002

CL, SB：GUESS?, Inc.

DOG DEPT **http://www.dogdept.com/** **JAPAN／2002**
CL, SB：ドッグデプト DOG DEPT DF：山田富男 Tomio Yamada

concept カリフォルニア州サンタモニカ発の、大好きな犬をモチーフにした自然派志向のカジュアルブランドです（マークにゴールデンレトリーバー犬をイメージしました。）Tシャツ、トレーナー、ポロシャツ、帽子、バック、ドッグ関連品、その他いろいろそろえています。

A futuristic design has been employed in keeping with the concept of high-performance underwear for future generations. Black, the corporate color, is used for a background that maximizes the sung fit of Under Armour wear. Streamlined, simple icons are used for navigation guides. The simplicity of this top page gives it the impact needed to draw the viewer further into the Flash site.

TODD OLDHAM JEANS **http://www.toddoldham.com** **USA／2002**
CL, SB：Todd Oldham Jeans

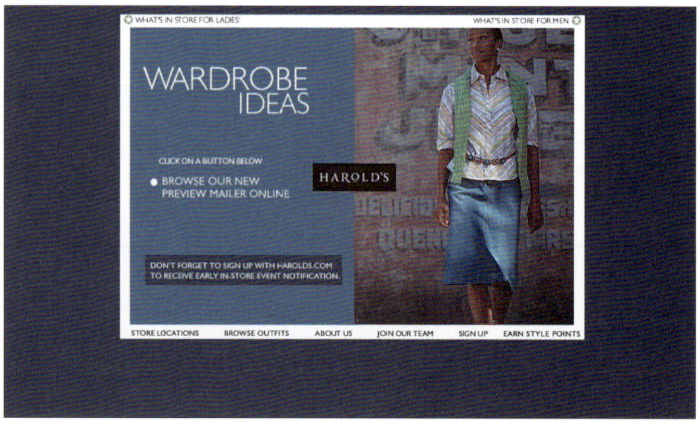

● **HAROLD'S** ● **http://www.harolds.com** ● **USA／2002**

CL, SB : Harold's Stores, Inc.

● **ROPÉ** ● **http://www.rope-jp.com**
● **JAPAN／2002**
CL, SB：(株)ジュン ロペ JUN CO., LTD. ROPÉ
DF：（株）フィーメール Fe-MAIL

concept ロペのコアである「上質な素材」「心地よさ」「ベーシックなデザイン」をそのままウェブデザインに置き換え、ホワイトスペースを大きくとり、「上質な緊張感」を持たせました。 メインビジュアル部分では大胆な遊びを考慮し、商品が引き立つレイアウトに。

ROPÉ's core concepts of quality, comfort, and basic design are incorporated in this website with a lot of white space that conveys a high-quality tension. The main visual applications highlight the appeal of the ROPÉ product line.

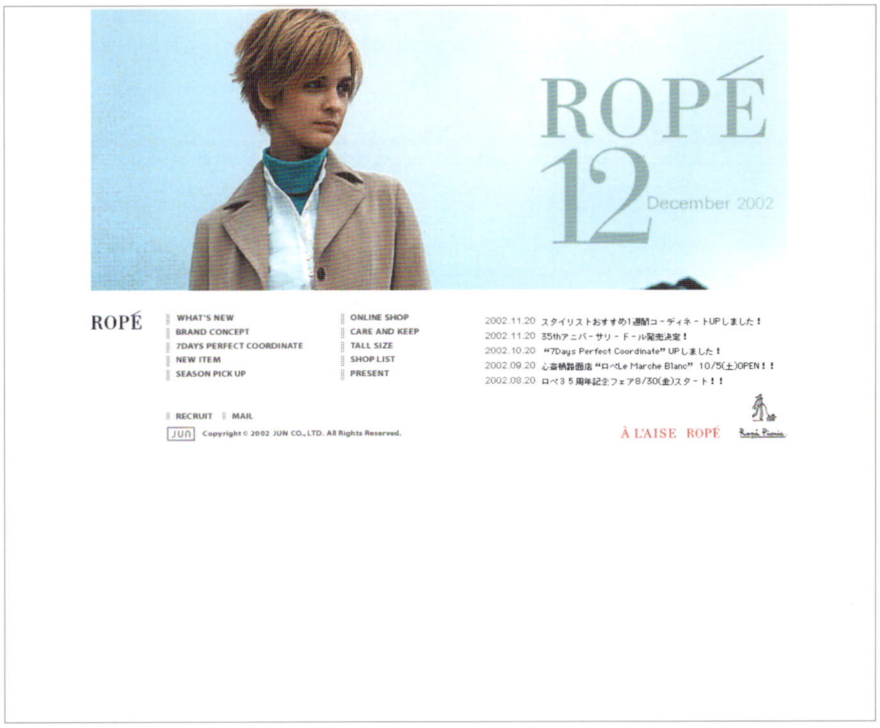

● **KAREN WALKER** ● **http://www.karenwalker.com** ● **NEW ZEALAND／2002**
CL, SB：KAREN WALKER DF：Alt Group

● **Natalia Naftalieva** ● **http://www.naftalieva.com** ● **RUSSIA**
CL, SB：Natalia Naftalieva

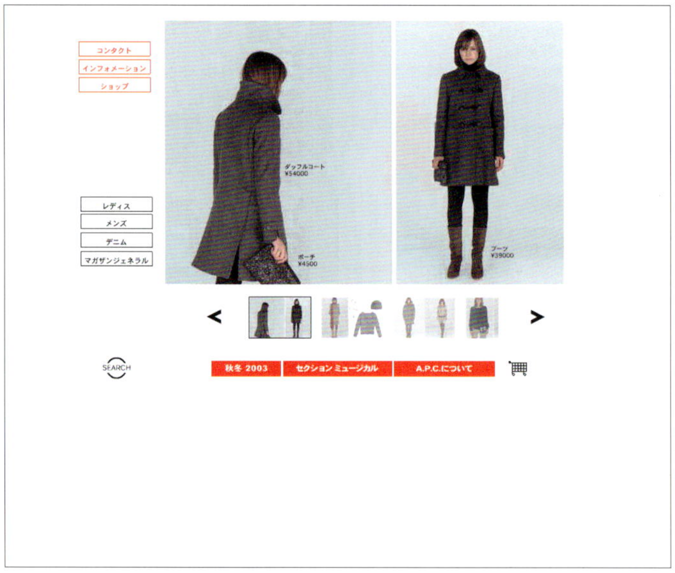

A.P.C. **http://www.apcjp.com/** **JAPAN／2002**

CL, SB：(株)イースト バイ ウエスト East by West Co., Ltd.

concept

ツールとしての最低限の機能とA.P.C.の世界観を、非常にミニマルなデザインの中に実現すること。

A minimalist approach has been take with this purely functional site in a reflection of the A.P.C. worldview.

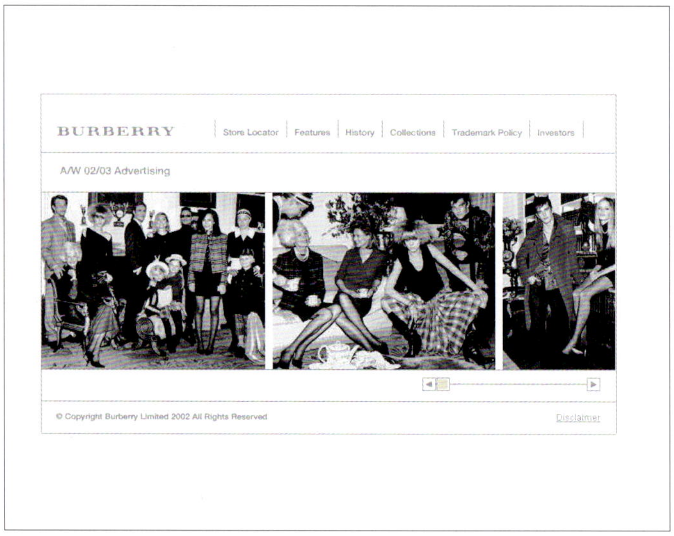

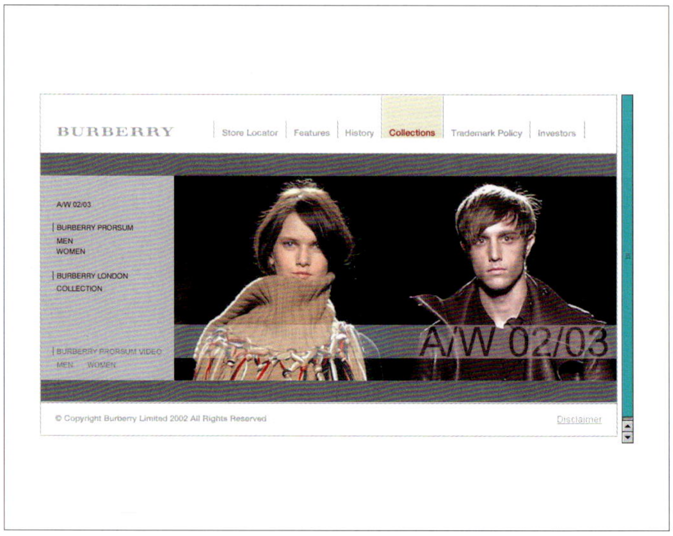

BURBERRY **http://www.burberry.com** **UK／2002**

CL, SB：Burberry Limited.

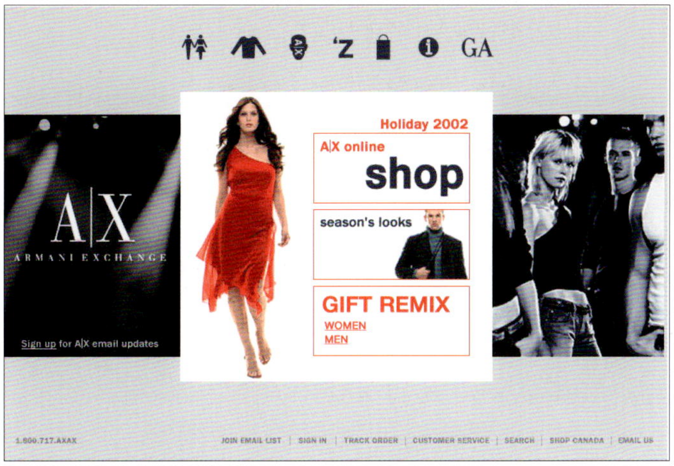

● **ARMANI EXCHANGE** ● **http://www.armaniexchange.com** ● **2002**
CL, SB：Armani Exchange

● **FRANCO MIRABELLI** ● **http://www.mirabelli.com** ● **CANADA／2002**
CL, SB：Franco Mirabelli Design

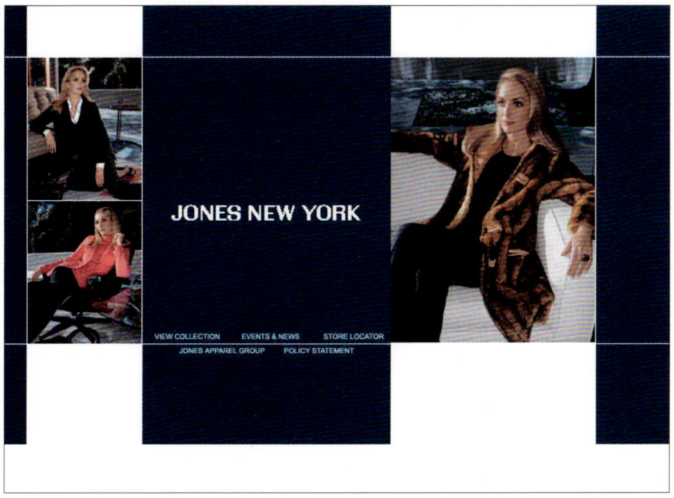

● **JONES NEW YORK** ● **http://www.jonesnewyork.com** ● **USA／2002**
CL, SB：Jones New York

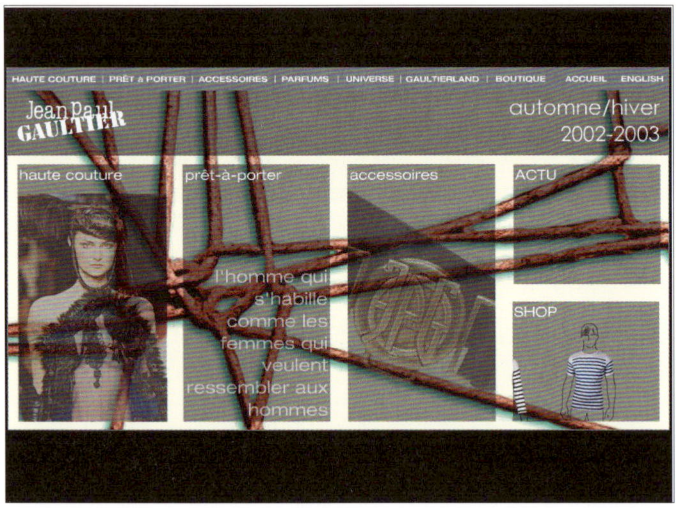

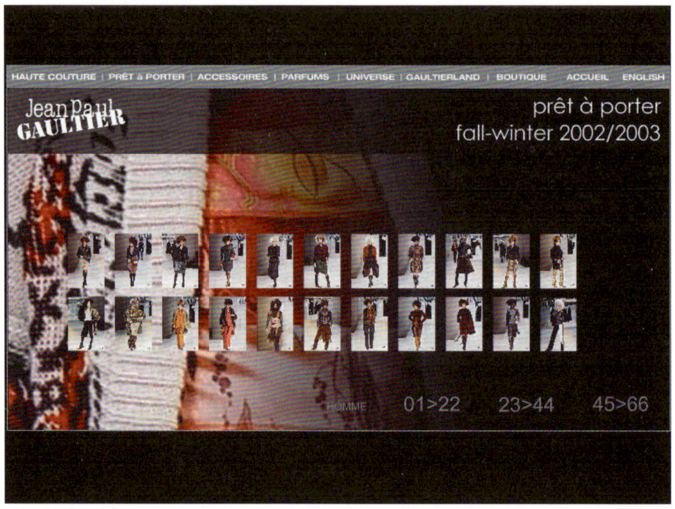

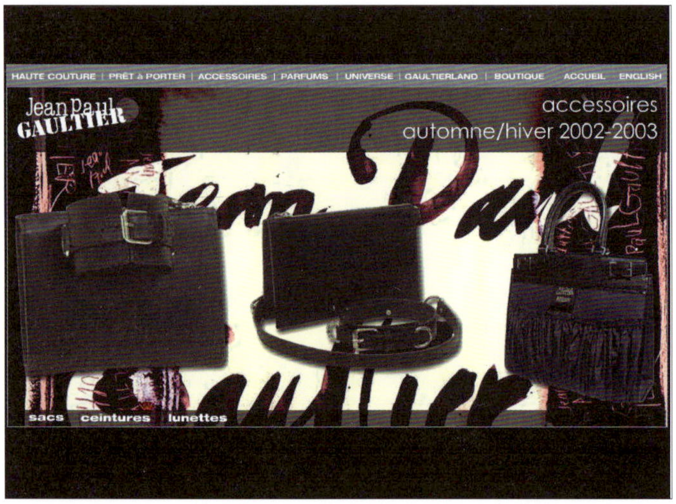

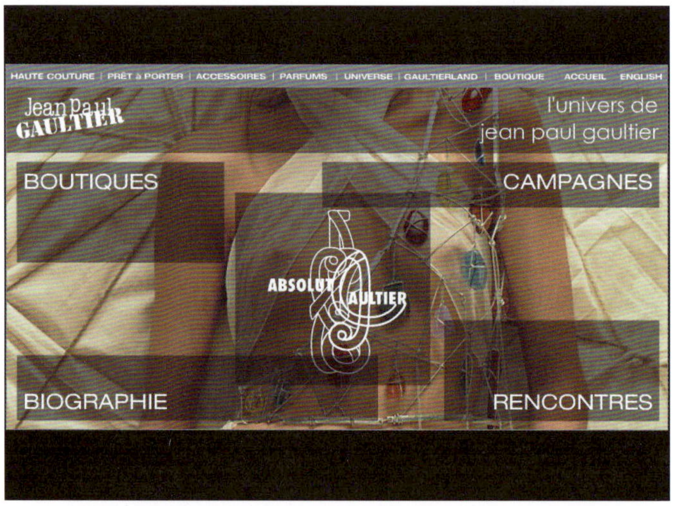

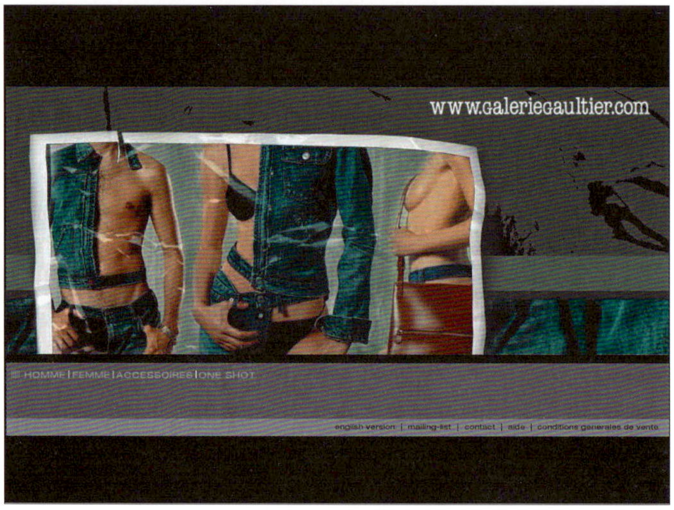

● **Jean Paul GAULTIER** ● **http://www.galeriegaultier.com** ● **FRANCE**
/2002 CL, SB : Jean Paul GAULTIE

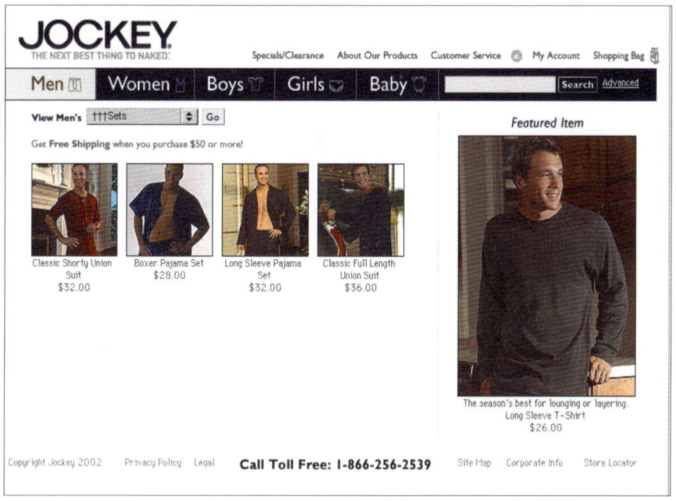

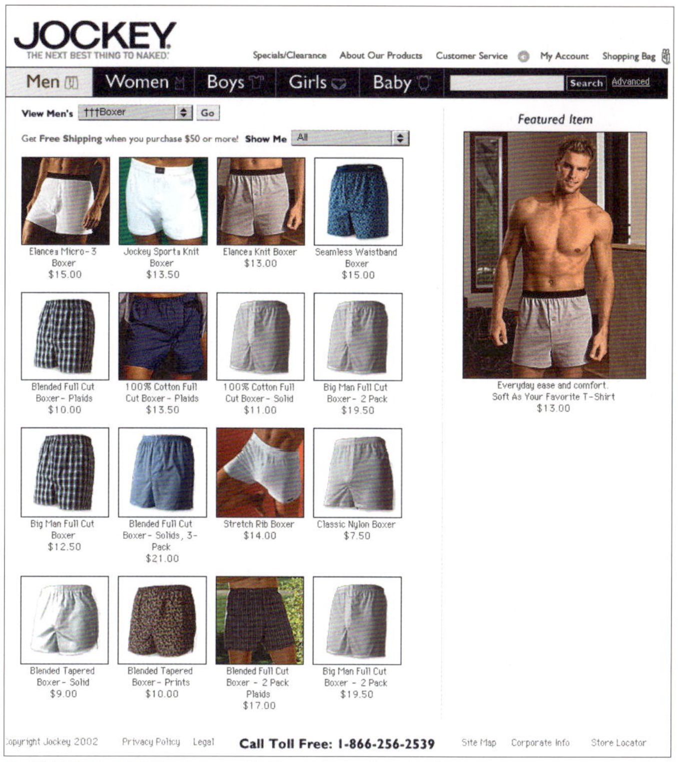

● JOCKEY ● http://www.jockey.com ● USA／2002

CL, SB：Jockey International, Inc.

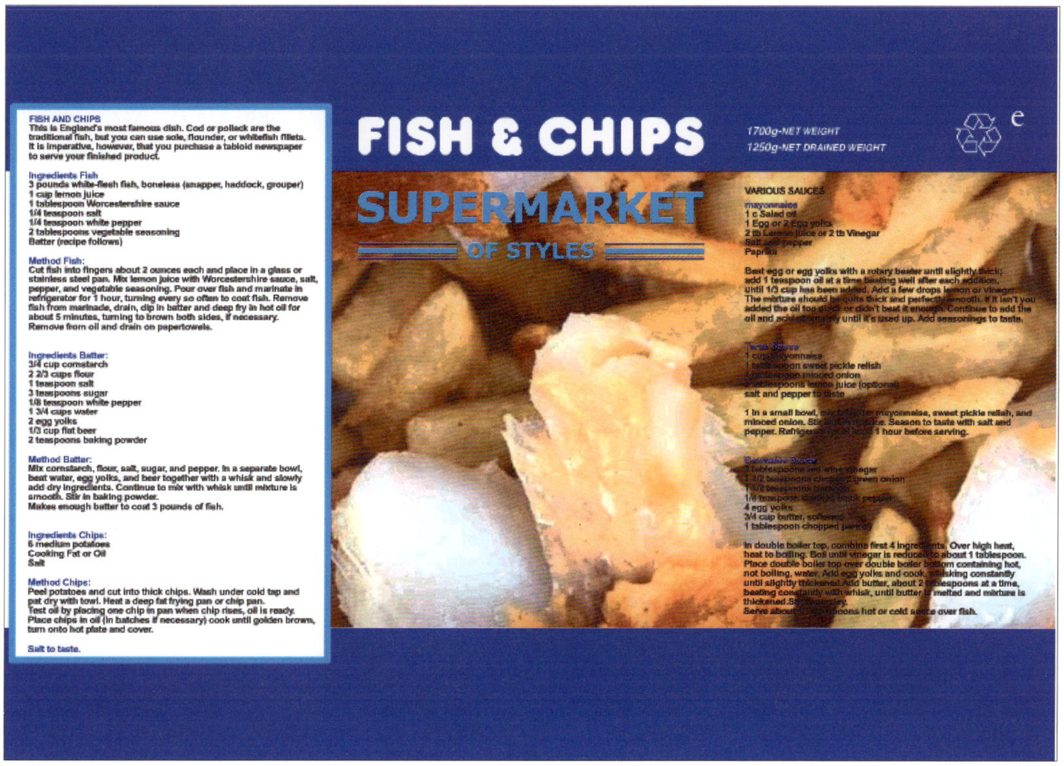

● FISH & CHIPS ● http://www.fishandchips.be ● USA／2002

CL, SB：FISH & CHIPS

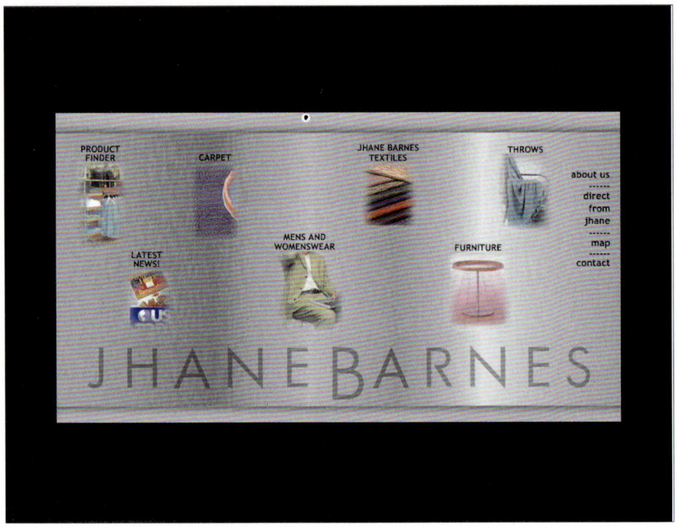

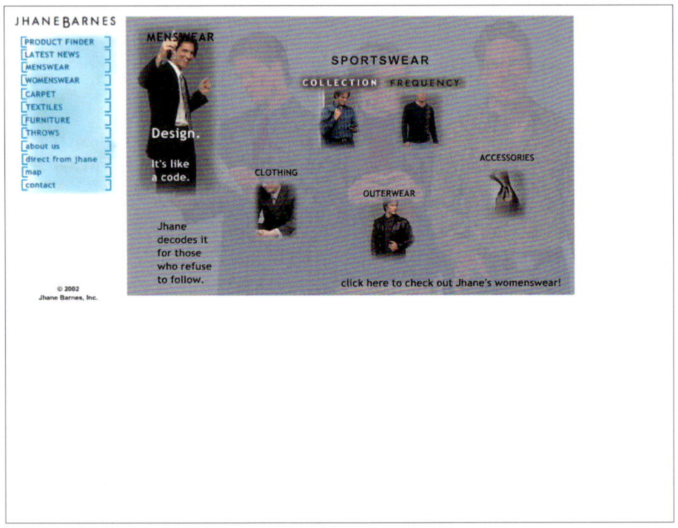

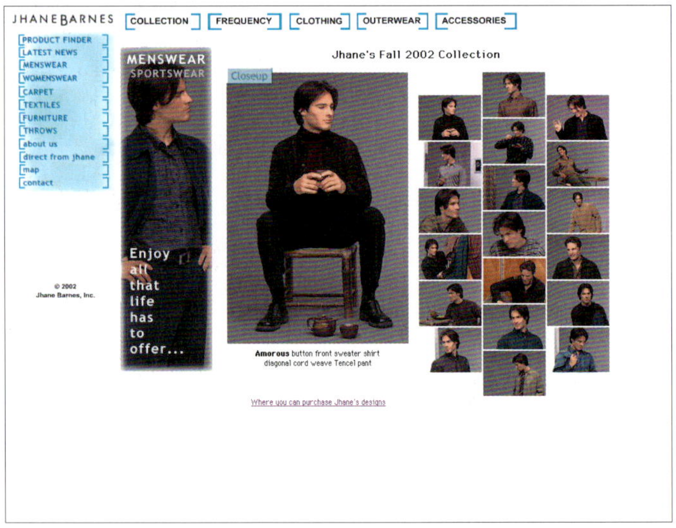

● JHANE BARNES ● http://www.jhanebarnes.com ● USA／2002
CL, SB : Jhane Barnes, Inc.

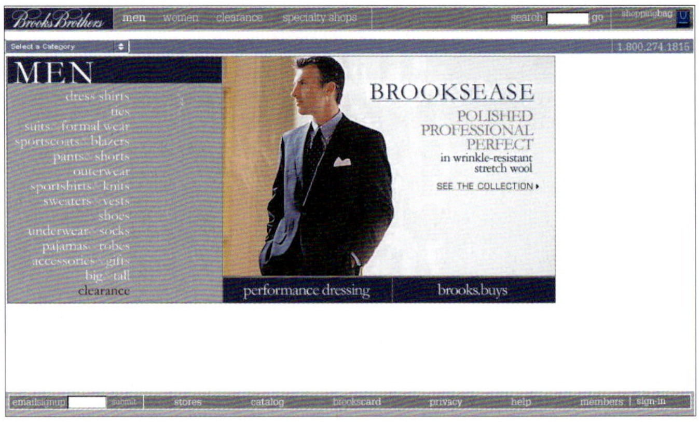

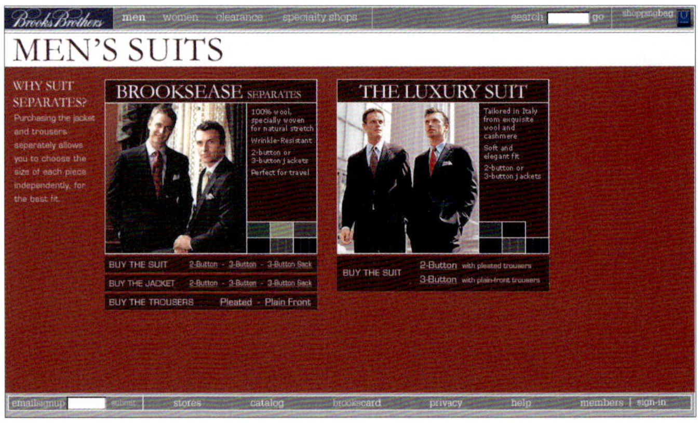

● Brooks Brothers ● http://www.brooksbrothers.com ● USA／2002
SB, DF : Brooks Brothers

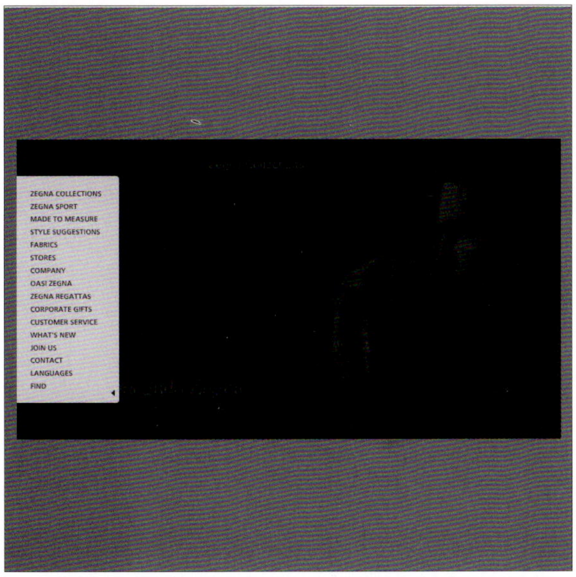

Ermenegildo Zegna　**http://www.zegna.com**　**ITALY／2002**
CL, SB：Ermenegildo Zegna

SHIPS WEB SITE　**http://www.shipsltd.co.jp**　**JAPAN／2002**
CL, SB, DF：（株）シップス SHIPS LTD.　DF：（株）オーセンティア Authentier, inc.

concept
“リアル＝店舗の販売支援のための使えるツール”としていくために、得たい情報をより得やすくすることを重視しました。

This site is intended to supplement store sales and as such emphasis has been placed on providing information in an easy-to-use format.

銀座テーラー GINZA TAILOR　**http://www.gintei.com**　**JAPAN／2002**
CL, SB：（株）銀座テーラー GINZA TAILORS' CO., LTD.　DF：鰐渕優子 Yuko Wanibuchi

● (株)ピー.ジー.エム P.G.M. CORP., LTD. ● http://www.p-g-m.co.jp
● JAPAN／2002
CL, SB：(株)ピー.ジー.エム P.G.M CORP., LTD. DF：(株)サイバースペース CYBER SPACE

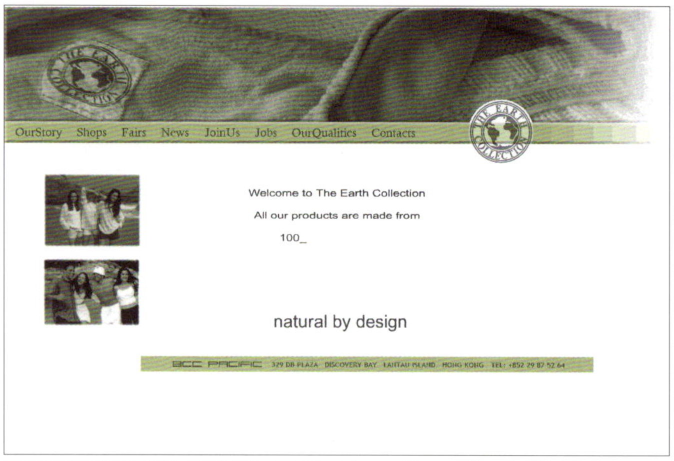

● THE EARTH COLLECTION ● http://www.theearthcollection.com
● CHINA／2002
CL, SB：THE EARTH COLLECTION

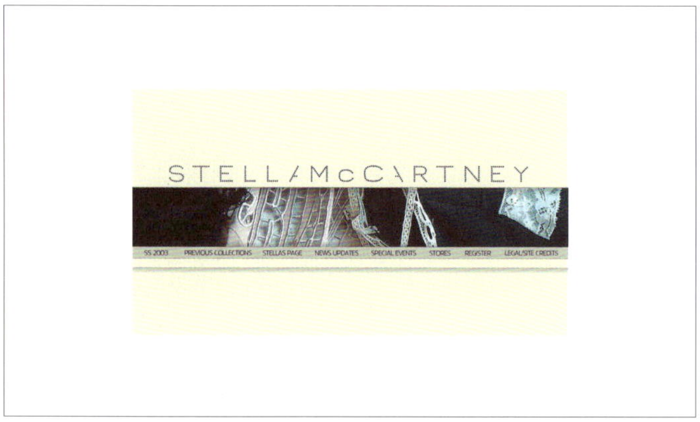

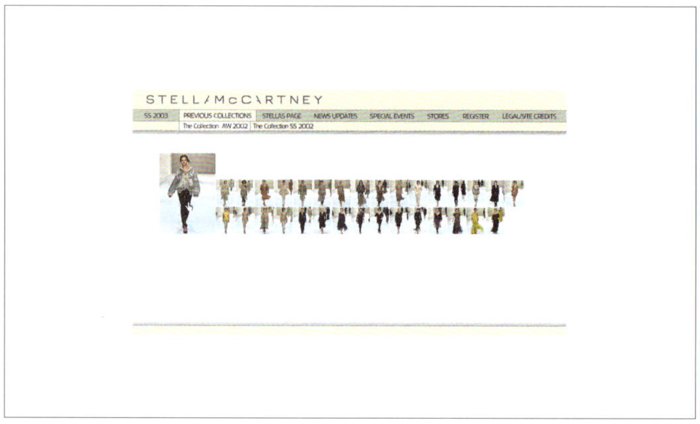

● STELLA McCARTNEY ● http://www.stellamccartney.com ● 2002
CL, SB：Gucci Group N.V. (Gucci)

● EMILIO PUCCI ● http://www.emiliopucci.com ● ITALY／2001
CL, SB：Emilio Pucci S.R.L.

● **CESARANI** ● **http://www.cesarani.com** ● **2002**
CL, SB：S.J.C. Concepts, Inc.

● **W.&L.T.** ● **http://www.walt.de** ● **GERMANY／2001**
CL：W.&L.T. fashion GmbH SB, DF：glimpse, berlin

concept 企業理念をコンテンツに反映させたWebサイト。試験的なEコマースサイトではなく、純粋な娯楽を体現するために作成しました。サイトに入ったらコンテンツをよく見て何かやってみてください。そうすれば、人生の意味と未来の広がりを感じていただけるはずです。

This is not a run-of-the-mill e-commerce site. It was designed to be sheer fun! Site design and content mirror the company's philosophy that you must interact to discover what life and the future offer.

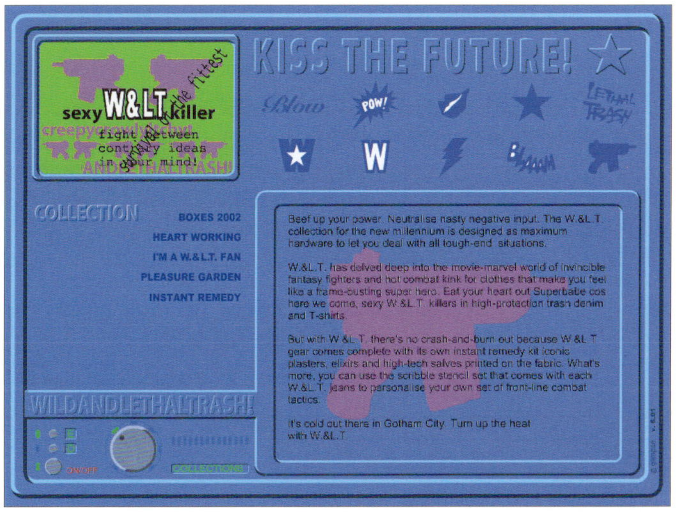

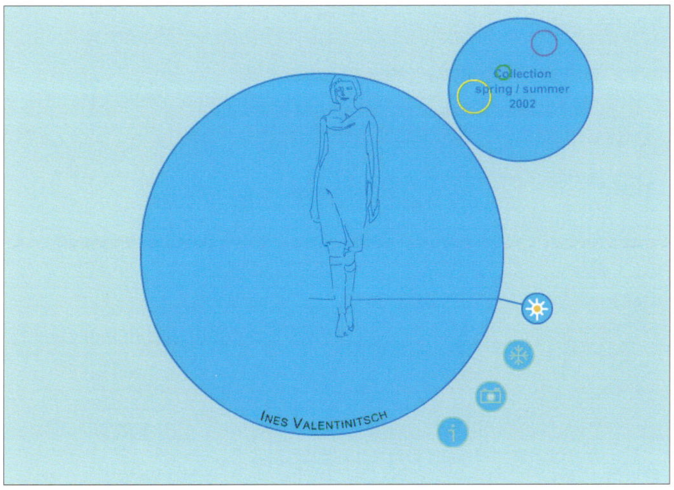

● **INES VALENTINITSCH** ● **http://www.inesvalentinitsch.com**
● **ITALY／2002**
CL, SB：Ines Valentinitsch

● **SAC** ● **http://www.sac-coltd.com** ● **JAPAN／2000**
CL, SB：(株)サック SAC Co., Ltd. DF：福井有正 Arimasa Fukui

● **LOWE VALENTINI** ● **http://www.loewevalentini.com** ● **ITALY／2002**
CL, SB：Lowe Valentini

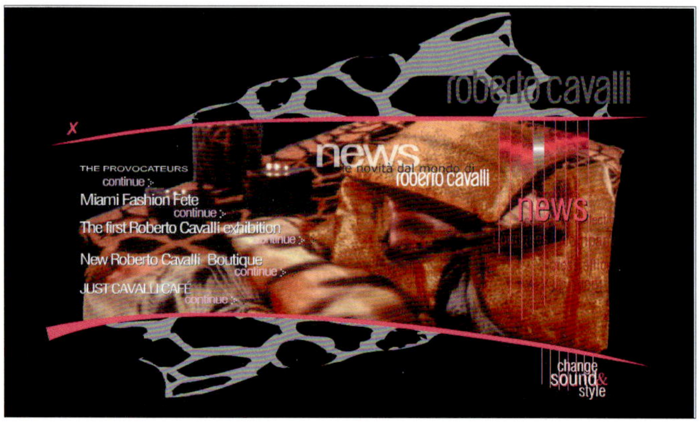

● **roberto cavalli** ● **http://www.robertocavalli.net** ● **ITALY／2002**
CL, SB：Roberto Cavalli　DF：zanigni

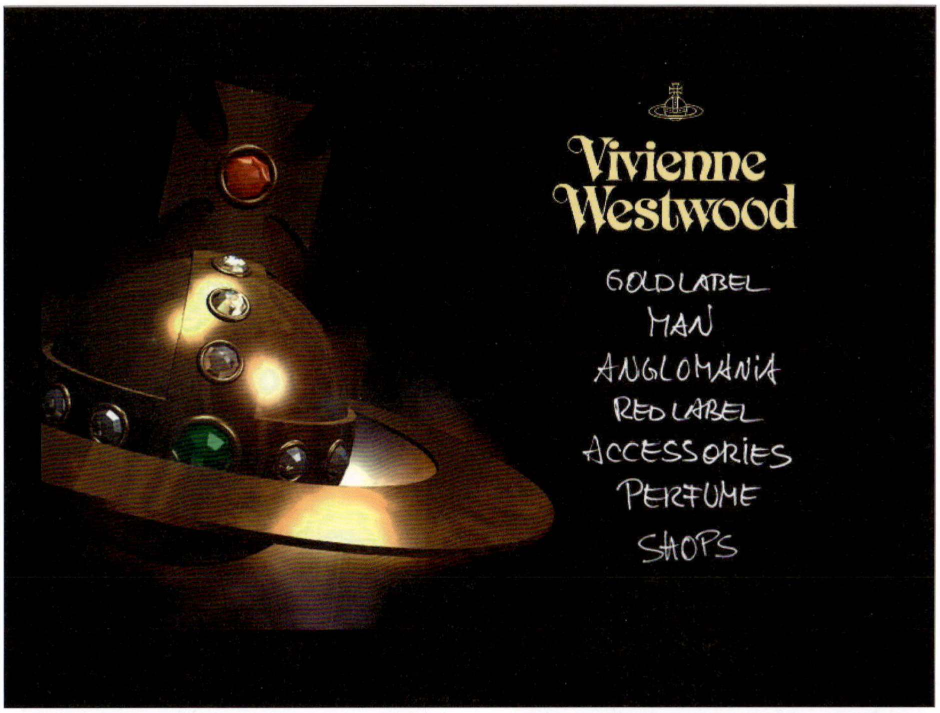

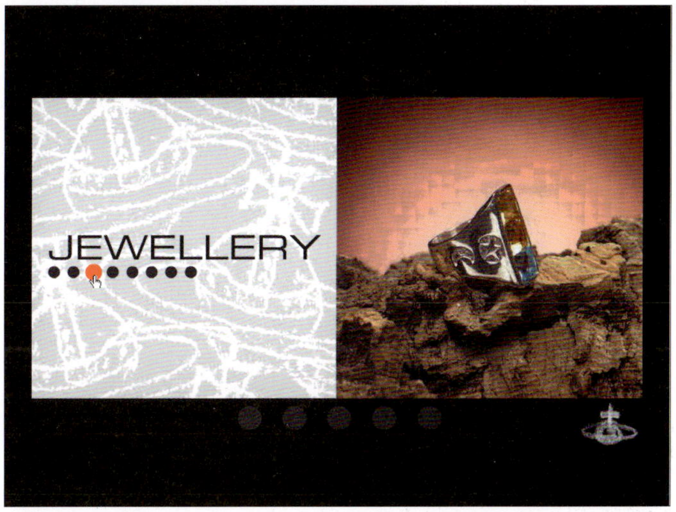

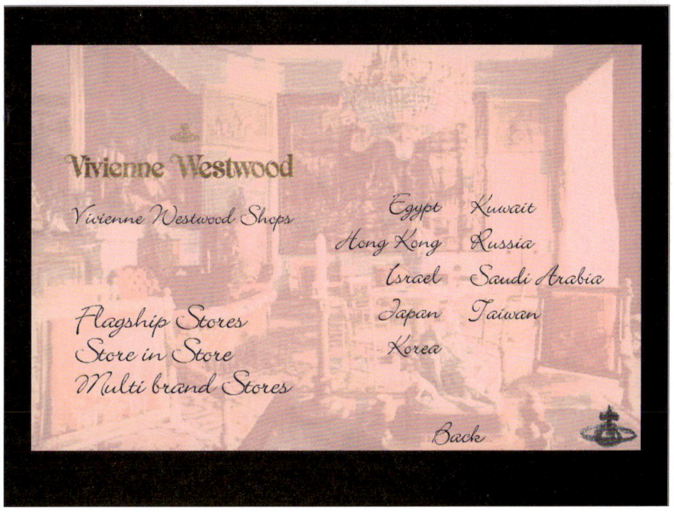

● **Vivienne Westwood** ● **http://www.viviennewestwood.com** ● **UK／2002**

CL, SB：Vivienne Westwood

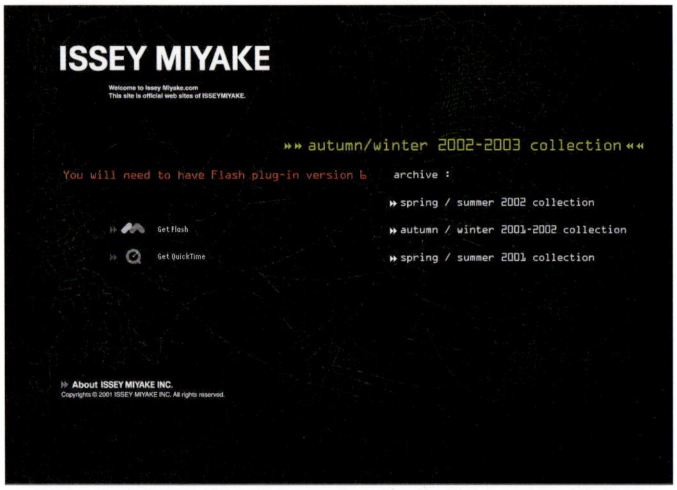

● **ISSEY MIYAKE** ● **http://www.isseymiyake.com** ● **FRANCE／2001**
CL：Issey Miyake SB：Incandescence DF：Etienne Mineur

concept 過去のコレクションを基にしたオンライン形式の「物語」で、イッセイ・ミヤケのメインデザイナーであるナオキ・タキザワの世界を展開し、コレクションをご覧に入れます。単なるスライドショーに収まることのない必見の価値あるサイトです。イッセイ・ミヤケのサウンドデザイナー、Sasha Gattinoの多大な尽力に感謝します。

The site was created to be a "narrative" that develops the universe of Naoki Takizawa (Issey Miyake's main designer) and shows the collections with a method that avoids becoming just a slide show. Special thanks to Sasha Gattino, Issey Miyake's sound designer.

● **YVES SAINT LAURENT** ● **http://www.ysl-hautecouture.com**
● **FRANCE／2001**
CL, SB：YSL HAUTE COUTURE

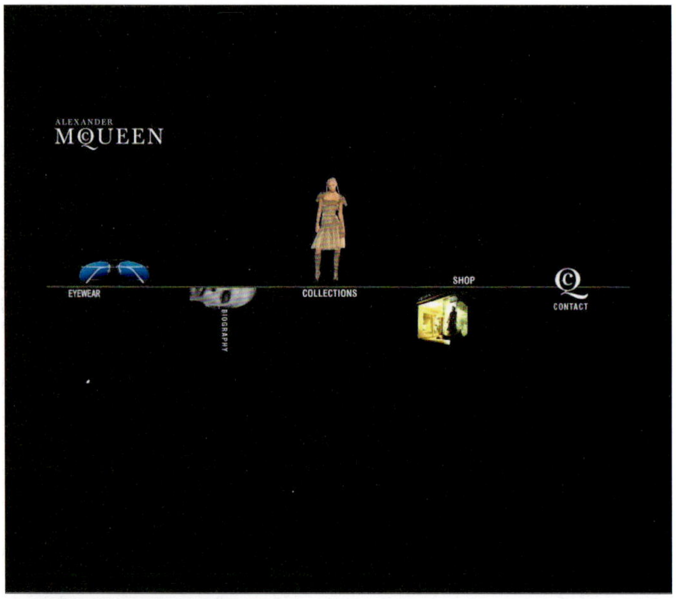

● **ALEXANDER McQUEEN** ● **http://www.alexandermcqueen.net**
● **UK／2002**
CL, SB：Alexander McQueen

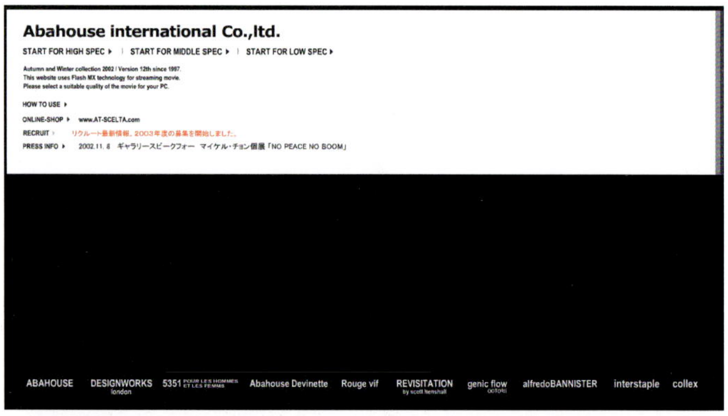

● **ABAHOUSE** ● **http://www.abahouse.co.jp** ● **JAPAN／2002**
CL, SB：(株)アバハウス インターナショナル Abahouse International Company DF：(有)フリフリカンパニー Furi Furi Company

● **SISLEY** ● **http://www.sisley.com** ● **ITALY／2002**
CL, SB：United Colors of Benetton

● **UNDER ARMOUR PERFORMANCE APPAREL**
● **http://www.underarmour.co.jp/** ● **JAPAN／2002**
CL, SB：(株)ドーム DOME Corporation DF：福田将之 Masayuki Fukuda

concept 「次世代のアンダーウェア」をコンセプトに未来感を感じさせるデザインにしました。コーポレートカラーである黒を背景にアンダーアーマーブランド商品の特徴であるタイト感をできるだけ大きく見せるようイメージを配置、ナビゲーション部分はシンプルなイメージアイコンで構成しました。フラッシュサイトのトップページとしてこの後の展開に興味を引かせるようインパクトのあるシンプルなデザインを心がけました。

A futuristic design has been employed in keeping with the concept of high-performance underwear for future generations. Black, the corporate color, is used for a background that maximizes the sung fit of Under Armour wear. Streamlined, simple icons are used for navigation guides. The simplicity of this top page gives it the impact needed to draw the viewer further into the Flash site.

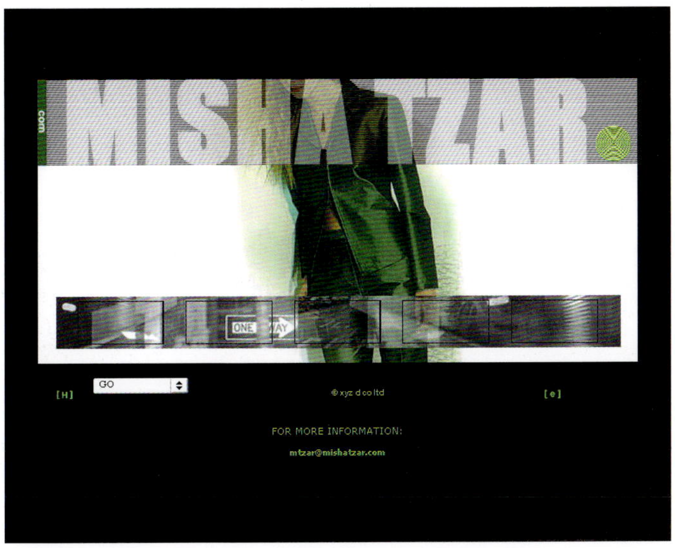

● **MISHA TZAR** ● **http://www.mishatzar.com** ● **USA／2002**
CL, SB：Misha Tzar / XYZ D. Co., Ltd.

● **ANNA SUI** ● **http://www.annasui.com** ● **USA／2002**
CL, SB：ANNA SUI

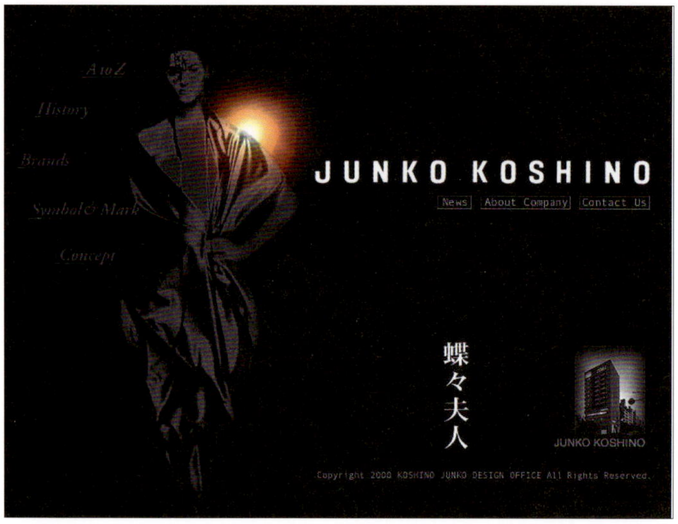

● JUNKO KOSHINO ● http://www.koshinojunko.com/ ● JAPAN／2000

CL, SB：(株)コシノジュンコデザインオフィス KOSHINO JUNKO DESIGN OFFICE DF：デザインアーク 若狭靖弘 DesignARC / Yasuhiro Wakasa

concept コシノジュンコデザインオフィスのコンセプト／イメージを表現。直接的にイメージを伝えるため、静止画での完成度を目指しました。ターゲットは、エンドユーザーではなく「クライアント」。近年のコシノジュンコのファッションにとどまらない幅広い活躍の場を基本にしています。各コーナーは、映画のスクリーンをモチーフにしたもので、SVGA以上の画面、ノンプラグインで破綻なく見ることができます。

Still pictures directly convey the image and concepts of the Junko Koshino Office. The target is better termed the "client" rather than the "end user." A movie screen motif employing SVGA-standard graphics is used to present in unbroken sequence the wide range of Junko Koshino fashion products.

● DKNY ● http://www.dkny.com ● USA／2002

CL, SB：The Donna Karan Company, DKNY

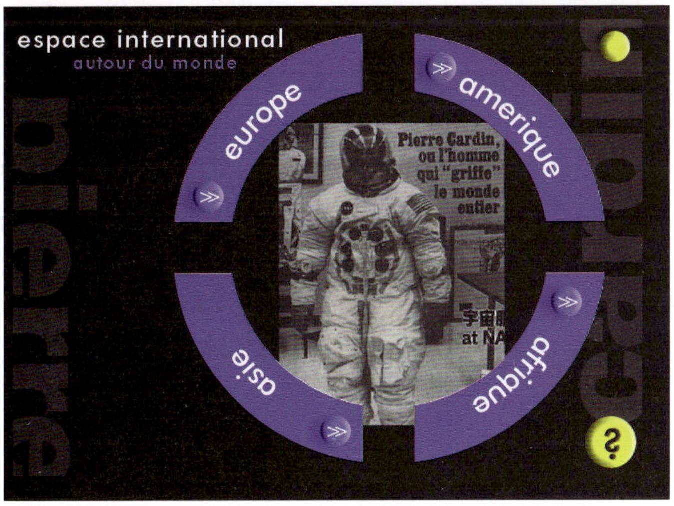

● pierre cardin ● http://www.pierrecardin.com ● FRANCE／2002
CL, SB：pierre cardin

● PIKO ● http://www.crymson.co.jp/piko/
● JAPAN／2002
CL：（株）クリムゾン Crymson Co.,Ltd. SB, DF：NTT ラーニングシステムズ（株） NTT LEARNING SYSTEMS CORPORATION

concept 『PIKO』とは、ネイティブハワイアンの古い語源で「おへそ」を意味します。サーフカジュアルブランド『PIKO』のページは、ハワイロコのイメージ、サーフィ ン、夏、海といった明るくスポーティーなイメージを生かしたFlashデザインページを制作しました。ハワイの素朴な風景と海のイメージの間に『PIKO』のロゴを挿入し、ロゴを印象付け るように仕上げています。

The brand name "Piko" is taken from the native Hawaiian word for navel. This Flash design page conveys the casual surfing brand's Hawaiian origins and reinforces the overall brand image by superimposing the Piko logo on local Hawaiian scenes.

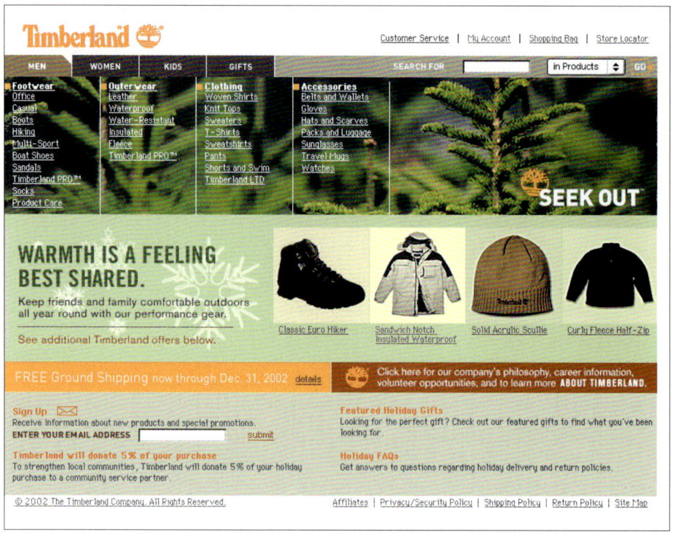

● Timberland ● http://www.timberland.com ● USA／2002
CL, SB：The Timberland Company.

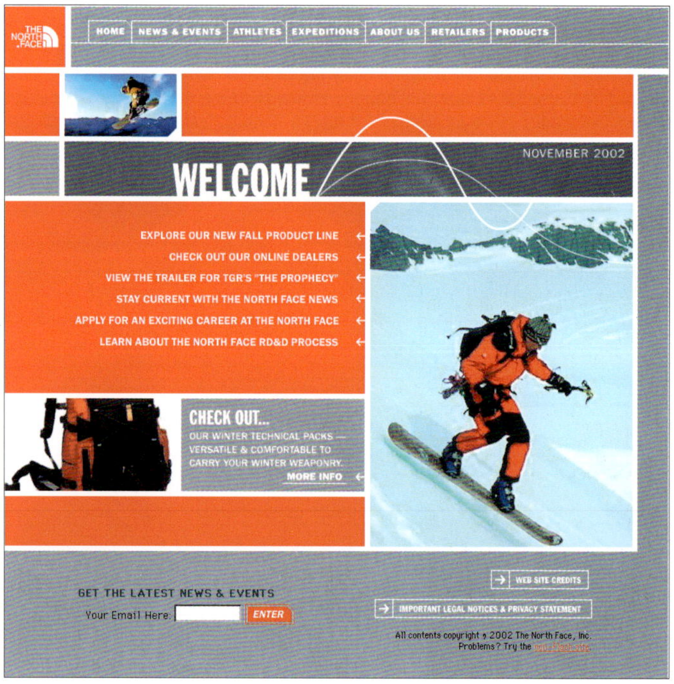

● THE NORTH FACE ● http://www.thenorthface.com ● USA／2002
CL, SB：The North Face, Inc.

● Carhartt ● http://www.carhartt.com ● USA／2002
CL, SB：Carhartt, Inc.

● ナイキジャパン（Nike Japan） ● http://www.nike.jp/women/top.htm ● JAPAN／2002 CL, SB：ナイキジャパン Nike Japan

● ナイキジャパン（Nike Japan） ● http://www.nike.jp ● JAPAN／2002
CL, SB：ナイキジャパン Nike Japan

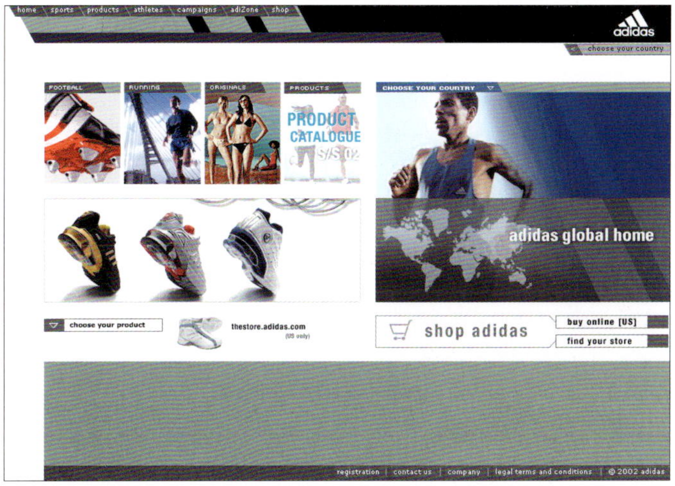

● adidas ● http://www.adidas.com ● GERMANY／2002
CL, SB：adidas-Salomon AG

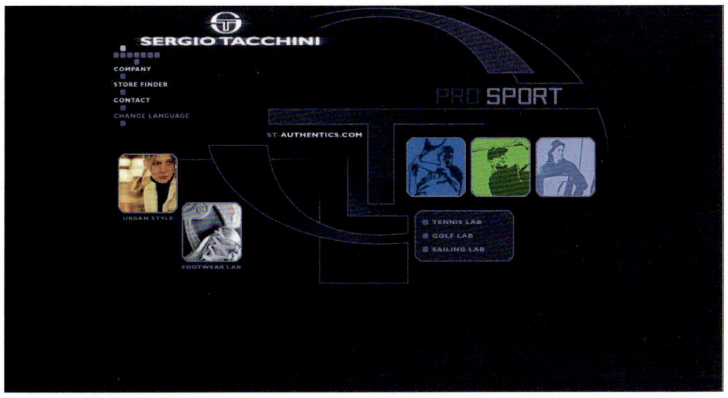

● SERGIO TACCHINI ● http://www.sergiotacchini.com ● ITALY／2002
CL, SB：SERGIO TACCHINI S.p.a.

● HELLY HANSEN ● http://www.hhworkwear.com ● NORWAY／2002
CL, SB：Helly Hansen Workwear

● PUMA ● http://www.puma.com ●GERMANY／2002
CL, SB：PUMA A.G.

● Robindira Unsworth ● http://www.robindira.com ● USA／2002
CL, SB：Robindira Unsworth

● Alohaloha ● http://www.alohaloha.com ● JAPAN／1999
CL：アロハロハ ALOHALOHA SB, DF：(株)スタイルワークショップ STYLE WORKSHOP INC.

● HUNTING WORLD ● http://www.huntingworld.co.jp/indexJ.html
● USA／2002
CL, SB：HUNTING WORLD

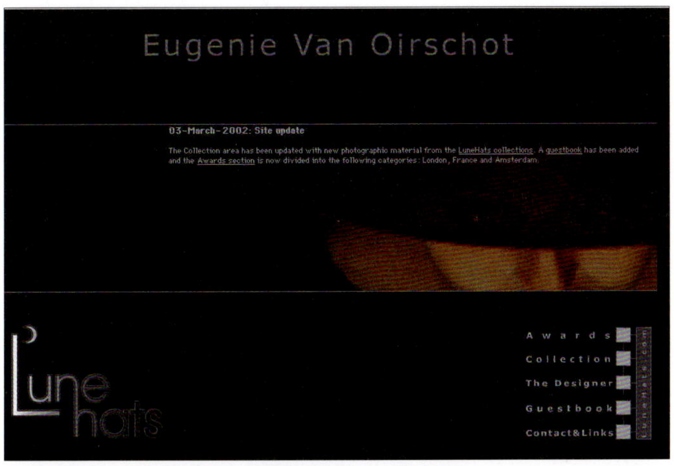

● Lunehats ● http://www.lunehats.nl ● UK／2002
CL, SB：Eugenie Van Oirschot-LuneHats

● シュウウエムラ（shu uemura）
● http://www.shu-uemura.com ● JAPAN／2002
CL, SB：（株）シュウウエムラ化粧品 shu uemura cosmetics inc.
DF：（株）トーキョウ・グレート・ヴィジュアル TOKYO GREAT VISUAL

concept

shu-uemura.comは、ブランド創始者のシュウウエムラの手の影がマウスのポインタの代わりとなり、アートの世界を体感できる、shu-uemura.co.jpのイントロダクション・サイト。2002年カンヌ国際広告賞 WEB・ブランドプロモーション部門で第3位（ブロンズ賞）を受賞しました。

The shadow of Shu Uemura's hand plays a major role in this introductory site to shu uemura.co. jp. This unique presentation was awarded the Bronze Cyber Lions Award for Brand Promotion at the 2002 Cannes International Advertising Festival.

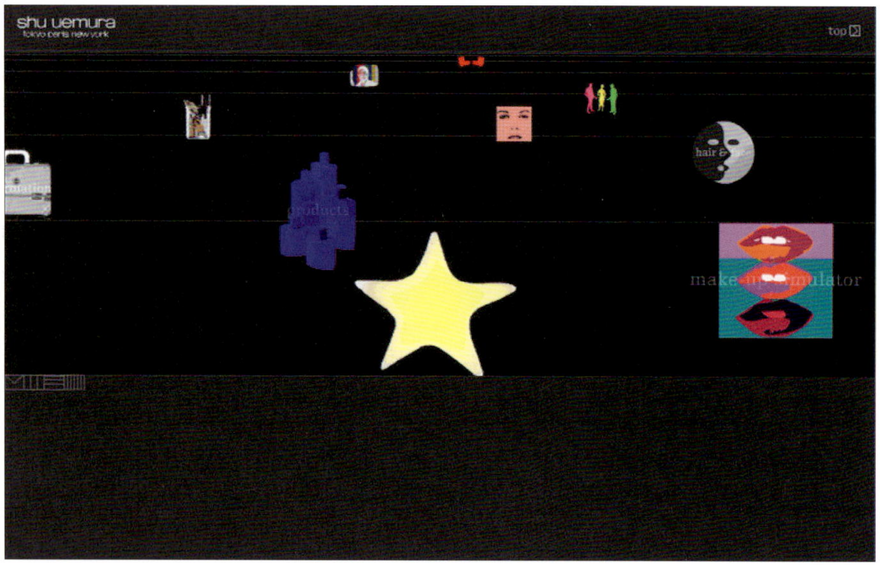

● http://www.shu-uemura.co.jp
CL, SB：（株）シュウウエムラ化粧品 shu uemura cosmetics inc.
DF：（株）トーキョウ・グレート・ヴィジュアル TOKYO GREAT VISUAL

concept

shu-uemura.co.jpは、テーマパークの中を散策するというコンセプトのもとに、スクリプト制御のフラッシュを多用し、ダイナミックなナビゲーションを実現しました。星の誕生を思わせるオープニング画面も、これ自体がローディング画面。続いて、espace(エスパース・宇宙)をモチーフとした空間に、アーティスティックなアイコンが点在し、各コンテンツの入口となっており、インタラクティビティを体感できるよう工夫。中でも「メイクアップ・シミュレーター」は、天文学的な数字のメイクアップのシミュレーションが可能なゲーム感覚のコンテンツで好評。

"Script-control" Flash is used for a dynamic navigation system providing the viewer with a theme-park like tour of the site. The Big Bang opening is actually a loading screen. This is followed by a space motif with artistic icons marking the entrances to the website's contents. The site is interactive, and the makeup simulator, in particular, offers an astronomical choice of makeup possibilities. This game-like approach has proved especially popular.

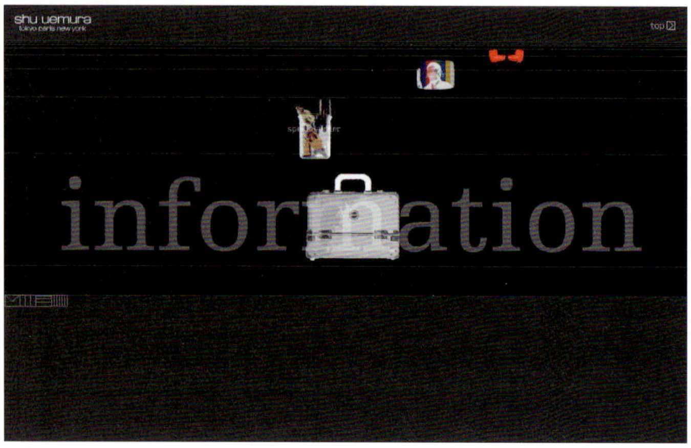

- **日本ロレアル（株）（NIHON L'ORÉAL K.K.)**
- **http://www.nihon-loreal.co.jp**
- **JAPAN／2002**

CL, SB：日本ロレアル（株） NIHON L'ORÉAL K.K.
DF：（株）イニット init co.,ltd.／Cythere (France)

concept 世界最大の化粧品会社ロレアルグループとその日本法人である日本ロレアル（株）の企業イメージを損なうことのないビジュアル、また主なターゲットとなるジャーナリスト、学生、就職希望者に最新の情報・ニュースをわかりやすい形で提供する事。

There were two objectives in designing this site. One was to use visual applications in keeping with the image of the world's largest cosmetics company as well as that of its Japan subsidiary. And the second was to present in an easy-to-understand format, information and news targeting journalists, students, and job seekers.

- **THE BODY SHOP**
- **http://www.bodyshop.com**
- **UK／2002**

CL, SB：THE BODY SHOP

● **AWAKE** ● **http://www.awake.co.jp/** ● **JAPAN／2001**
CL：(株)コーセー KOSÉ Corporation. SB, DF：(株)ブロードテック BROADTECH INC.

concept 都市生活者を中心にした女性をターゲットに、Innovative technology & styleを提唱する化粧品ブランド「アウェイク」。洗練されたブランドイメージを損ねることなく、ユーザビリティの高いサイトデザインを追求。ブランドアイデンティティとテクノロジー、コミュニケーションスタイルプロダクツ情報、販売店情報などを掲載。

Targeted at the cosmopolitan woman, AWAKE is a line of cosmetics distinguished by its innovative technology and style. The website retains the sophisticated brand image of the product line while at the same time providing easy access to information on brand identity and technology as well as individual products and shops.

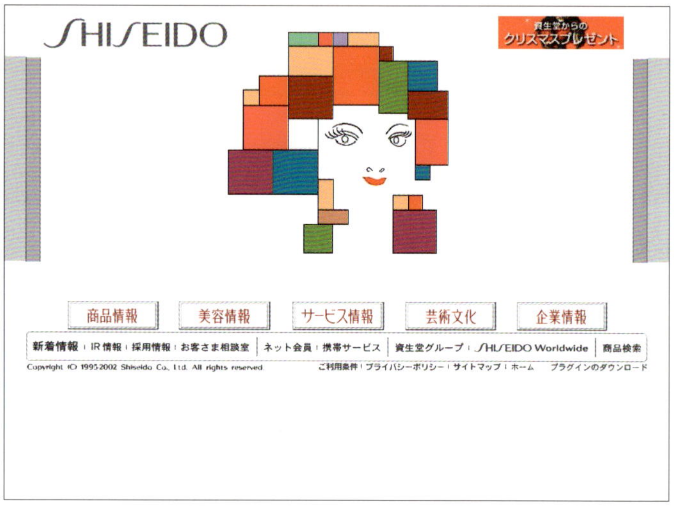

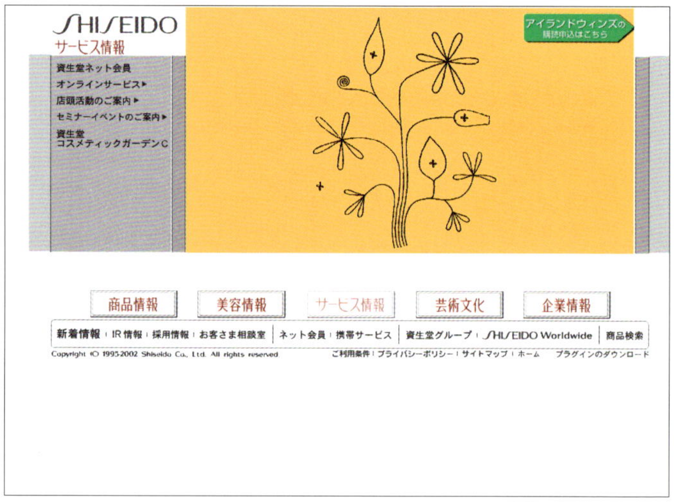

● **資生堂(SHISEIDO)** ● **http://www.shiseido.co.jp** ● **JAPAN／2002**
CL, SB, DF：(株)資生堂 Shiseido Co., Ltd.

concept インターネットの普及にともないお客さまとの重要なコミュニケーションツールとなった資生堂ホームページ「資生堂ウェブサイト(旧呼称：サイバーアイランド)」を、2002年5月、全社的な視点から戦略的に見直し、リニューアルしました。1995年開設当時より強みであった「デザイン力の強さ」を活かしつつ、会員の9割が女性という特長を踏まえ、シンプルでありながらも気品と伝統を伝えるデザインを目指しました。また、発信情報については資生堂グループ全体の顔として、ここにくればオール資生堂の情報にたどり着けるように企画・構築しました。具体的には、創業130周年を踏まえ山名文夫氏のイラストを用い資生堂らしさを表現するとともに、発信されている膨大な情報をお客さま視点にたち、「商品情報」「美容情報」「企業情報」などに再整理をし、情報を探しやすくしました。また、基本的にはどのコンテンツにもトップページから2クリックでたどり着けるようになっています。この秋には、海外を含むグローバルトップであるshiseido.comのリニューアルも予定しており、更に使いやすく、魅力的なホームページとなるよう進化させていきますので、ご注目ください。

With the widespread dissemination of the Internet, the Shiseido website (formerly known as the Shiseido Cyber Island) underwent a complete renewal in 2002 May in line with a company-wide review of corporate strategies. Design has always been the site's strength since it first opened in 1995, but this time a special effort was made to create a simple yet sophisticated format directed at a primarily female audience that would still convey Shiseido's long and proud traditions. The site is intended to give a comprehensive overview of all Shiseido's activities. Ayao Yamana's illustrations have been used in commemoration of the 130 years since Shiseido's founding, and information has been sub-divided into customer-friendly and easy-to understand categories such as "Product Information," "Beauty Information," "Corporate Information," and the like. Every section is just two clicks away from the top page.

FOOD

Cafe & Restaurant
[カフェ&レストラン]

Food & Drink
[食品&飲料]

Confectionery
[菓子]

● モスバーガー(MOS BURGER) ● http://www.mos.co.jp ● JAPAN／1996
CL, SB： (株) モスフードサービス MOS FOOD SERVICES, INC.

concept モスバーガー店舗で体験できる「おいしさ」や「くつろぎ」をそのまま再現。特に商品画像の制度には力を入れています。全体の統一感を保ちつつ、サイト構成を簡略化し、アクセスのしやすさにも配慮しています。

Our objective was to reproduce the deliciousness and relaxation customers experience at Mos Burger stores. In particular, we concentrated on detail in the product images. We gave the site a uniform look and simplified the structure to facilitate navigation.

● **養老乃瀧（YORONOTAKI）**　● **http://www.yoronotaki.co.jp**　● **Japan／2001**

CL, SB：養老乃瀧（株） YORONOTAKI Co., Ltd.　DF：日本システムウエア（株） NIPPON SYSTEMWARE CO.,LTD. (Total Page Design)
（株）マーヴィンズ Marvin's Inc. (Official Page Design)

concept 弊社はフランチャイズチェーンの為、加盟希望者が見て弊社を理解しやすい様にという部分と、一般のお客様が楽しんで頂ける様にという部分で総合トップからの選択制にしました。デザインコンセプトは、全体的にシンプルでアットホームな雰囲気。飲食業なので、清潔感が持てるようなHPを心がけました。

YORONOTAKI is a franchise chain, so the site's top page is in a table-of-contents format that enables easy navigation to two sets of pages. One set presents company information in a clear and intelligible way for visitors interested in the franchise opportunity; the other set is entertaining content for general visitors. The simple design projects an at-home feel and a clean look appropriate for a food and beverage business.

● **ラ・カシータ(La Casita)**　● **http://www.lacasita.co.jp/**　● **JAPAN／2002**

CL, SB：（有）ラ・カシータ La Casita　DF：ザ キュア ルーム the cure room

● **Chef Paul Prudhomme**　● **http://www.chefpaul.com/**　● **USA／2002**

CL, SB：Chef Paul Prudhomme

● Ciao! bambina
● http://www.ne.jp/asahi/ciao/bambina
● JAPAN／2000
CL, SB, DF：島村由花　Yuka Shimamura

concept 『カフェ』をキーワードにどれだけの人がつながるのだろうか。ひとそれぞれがもつ『カフェ観』を少しずつ紹介し、見る人自身も自分が好きなカフェを見つけるきっかけになればいいと思い立ち上げました。

We wanted to see how many people we could get together at a site built on the theme "cafe." Site users write about what they like about their favorite cafes and how cafes fit into their lifestyle. These mini-essays are introduced a few at a time to inspire site visitors to discover their own favorite cafes.

● Deco's Dog Cafe
● http://www.hot-dog.co.jp/
● JAPAN／2002
CL, SB：(株) ホットドッグ　Hot Dog co., ltd.

● モトヤエクスプレス (MOTOYA EXPRESS)
● http://www.motoya-exp.com　● JAPAN／2001
CL, SB：(有) 伊藤素樹事務所　Motoki Ito Office, Inc.　DF：グレート・ローク・アソシエイツ　GREAT LOG ASSOCIATES CO., LTD.

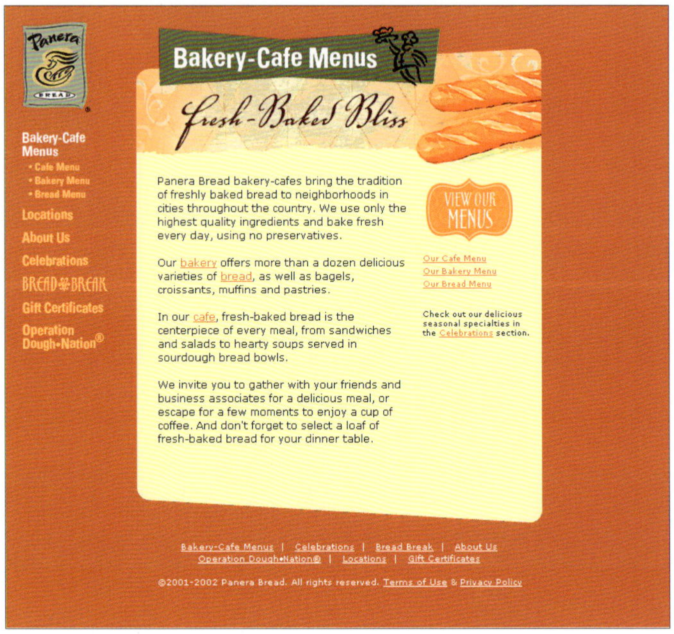

● Panera Bread　● http://www.panera.com/　● USA／2002
CL, SB：Panera Bread

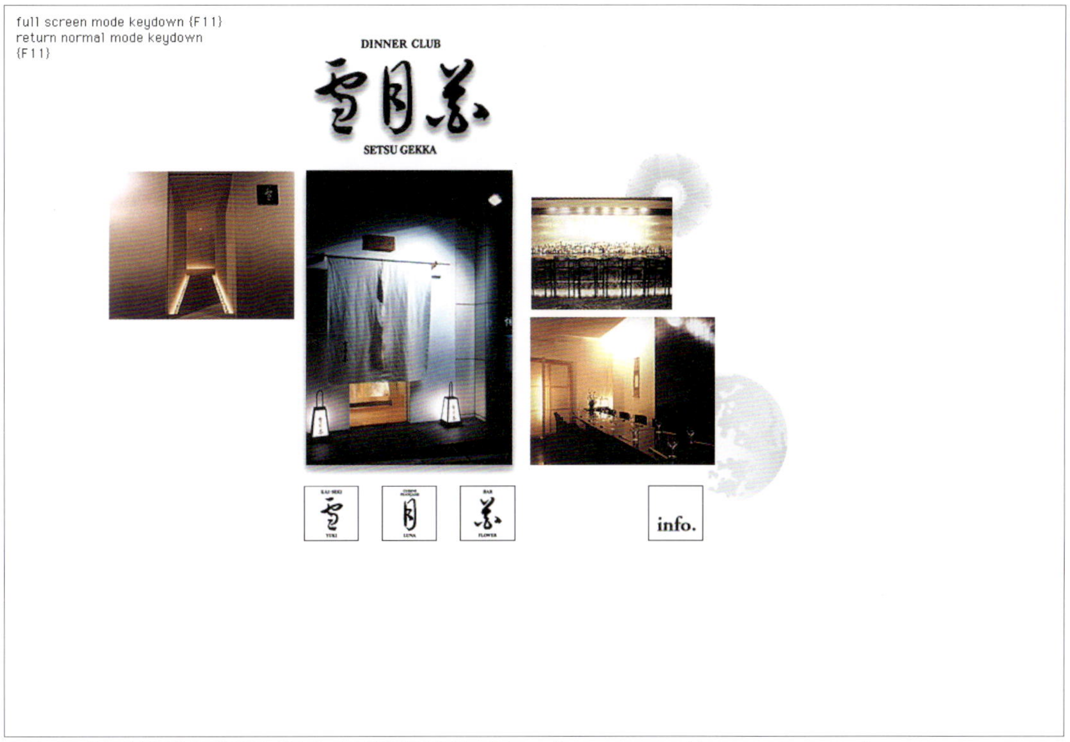

● ディナークラブ 雪月花(DINNER CLUB SETSUGEKKA) ● http://www.setsugekka.com ● JAPAN／1999
CL, SB：ディナークラブ 雪月花 DINNER CLUB SETSUGEKKA DF：後藤守男 Morio Gotou

● 鳥良(TORIYOSHI) ● http://www.toriyoshi.co.jp ● JAPAN／2002
CL, SB：サムカワフードプランニング(株) SAMUKAWA Food Planning Co., Ltd. DF：(株)NKB NKB INC.

● キッコーマン(KIKKOMAN) ● http://www.kikkoman.co.jp/ ● JAPAN／2002
CL, SB：キッコーマン(株) Kikkoman Corporation.

● オハヨー乳業 (OHAYO)
● http://www.ohayo-milk.co.jp/ ● JAPAN／2002
CL：オハヨー乳業(株) OHAYO DAIRY PRODUCTS CO., LTD. SB, DF：(株)オフィスプルーフ Office PROOF

concept 乳製品メーカーとしてふさわしいように、全体のデザインをさわやかなトーンにし、商品紹介だけでなく関連情報を充実させ、メーカーとしてのサービス向上を図りました。

The overall design creates a refreshing effect that harmonizes with the image of a diary products manufacturer. The site doesn't limit itself to simple product introduction; we've included a broad range of information on related topics for the benefit and enjoyment of visitors to the site.

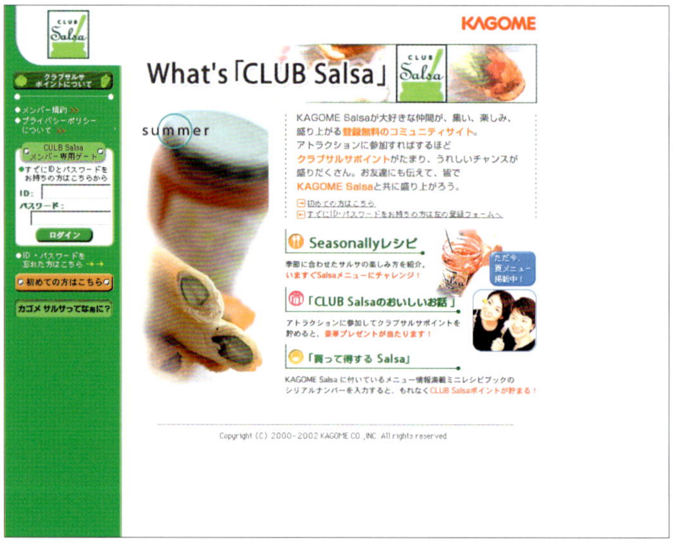

● KAGOME ● http://www.club-salsa.com ● JAPAN／2001
CL, SB：カゴメ(株) KAGOME CO., INC. DF：(株)大広 DAIKO ADVERTISING INC.

concept 食を楽しくオシャレに演出してくれるカゴメサルサを中心に同世代の女性コミュニティサイトとしてOPENしました。サイト内では発言や参加アトラクションにポイントを付与し常に活気を保っています。レシピ提案を中心とした中で香辛料の如くピリリと効いたサイトにしています。

The site was opened to build up a community of women fans of Kagome Salsa, a product that adds fun and pizzazz to meals. Users register as Club Salsa members and receive points for making suggestions and participating in various events, so the site is always hopping with activity. Recipe suggestions play a central role. We try to make the site as spicy and effective as our product.

● 伊藤ハム（Ito Ham Foods）　● http://itoham.mediagalaxy.ne.jp/
● JAPAN／2002
CL, SB：伊藤ハム（株）　ITOHAM FOODS Inc.　DF：DNPデジタルコム　DNP DIGITALCOM CO.,LTD.

concept 商品のメイン購買層である20代～40代の女性を強く意識して構成しています。今回のトップページは2002年11月に全面リニューアルした作品で、アクセス数の非常に多いレシピを全面に出して強調しているほか、コーナー説明も簡素化し、必要な情報はポップアップするなどユーザビリティを強化しています。また、バナーの数を減らし、全体的にすっきりしたトップに仕上げました。

This site is directed at the main consumer cohort for Ito Ham products, women in their 20s to 40s. The emphasis of the current top page, renewed in November 2002, is on recipes that have proven extremely popular, and simplified topics with pop-up explanations for enhanced usability. The number of banners has been cut back as well for an overall sleek and refined image.

● S&B　● http://www.torokeru.com/
● JAPAN／2000
CL, SB：エスビー食品（株）　S&B FOODS INC.　DF：（株）アサツー ディ・ケイ　ASATSU-DK INC.

concept 「秋の味覚でキレイになろう！」、「風邪の予防対策は大丈夫？」、「春のストレス解消法」など家族の健康と食生活にスポットをあてて、ホットでとろけるような生活情報提供コンテンツと、家庭で手軽にとろける気分と味わいが楽しめるレシピ紹介ページを常設。季節やキャンペーンに連動して、暮らしに役立つ栄養知識がゲーム感覚で学べるクイズコーナーや、日常生活のとろける度を診断する簡単なチェックコーナー、アンケートなどのコンテンツを設け、楽しさ山盛りで紹介しています。

This is an informational site that spotlights family health and dietary habits with articles on interesting topics such as "How the Tastes of Autumn Can Make You Beautiful!", "Are You Taking Adequate Cold-Prevention Measures?", and "Cures for Stress in Spring." A permanent recipe page features easy and delicious recipes made with Torokeru sauce mixes. Quizzes connected with the changing seasons and campaigns let users take a game approach to learning about nutrition. The site has lots of other fun pages like "Diagnose Your Torokeru Level" and questionnaire results.

● キッコーマン（KIKKOMAN）
● http://www.kikkoman.co.jp/homecook/　● JAPAN／2002
CL, SB：キッコーマン（株）　Kikkoman Corporation.

● なとり（Natori）　● http://www.natori.co.jp/　● JAPAN／2000
CL：（株）なとり　NATORI CO.,LTD.　SB, DF：（株）CSネット21　CSnet21 Co.,Ltd.

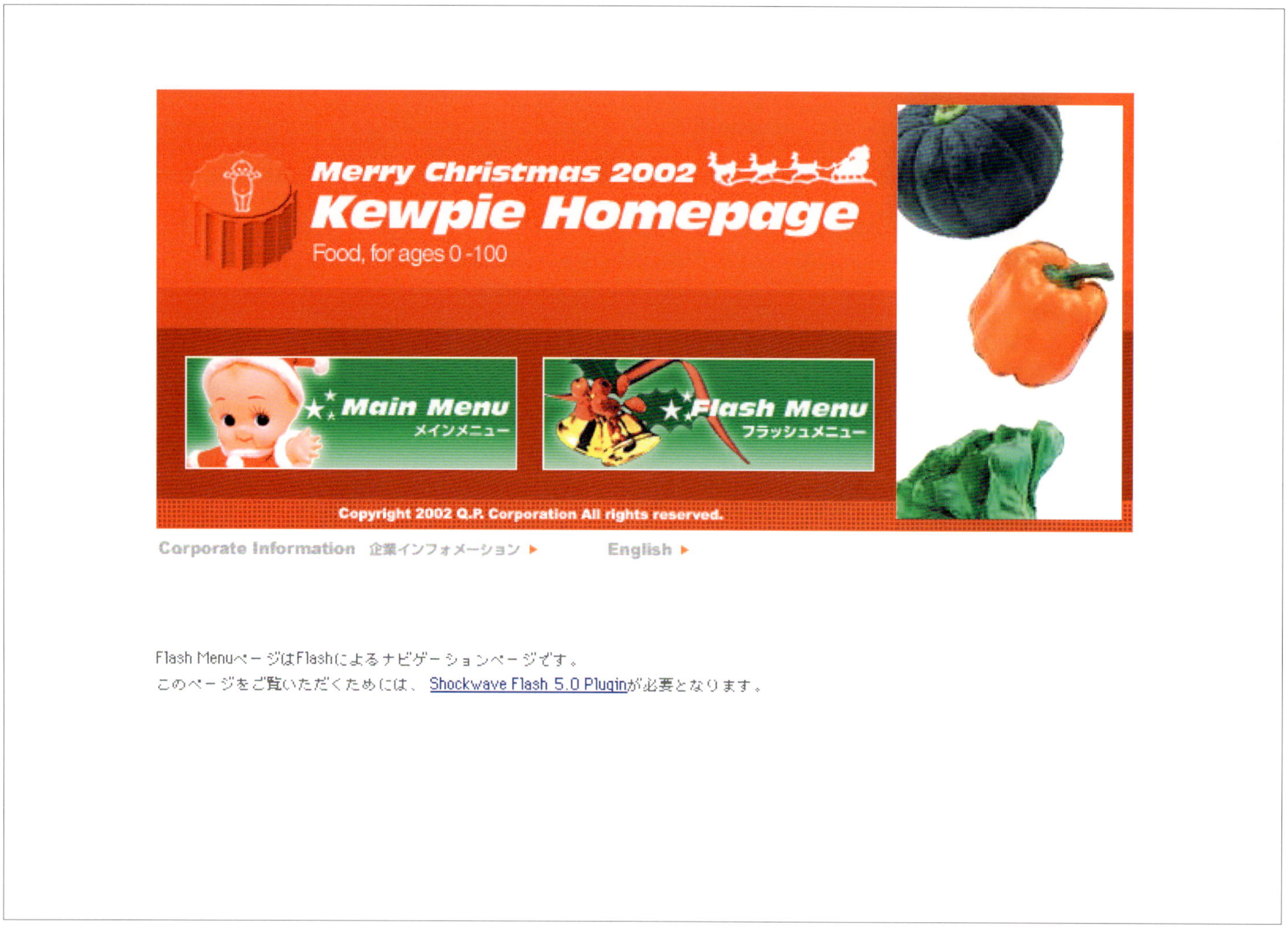

● キユーピー(Kewpie) ● http://www.kewpie.co.jp/ ● JAPAN／2001
CL, SB：キユーピー(株) Q.P. Corporation DF：(株)朝日広告社 Asahi Advertising Inc.

concept 若年主婦を対象にした、子供と一緒に楽しめる、簡単で楽しく検索しやすいサイトをデザインしています。操作性を簡単にしたレシピ検索「とっておきレシピ」では、検索レシピのほかに、今週のレシピや、マヨネーズ、ドレッシングを使ったレシピなどに分けて紹介。また、「キユーピーキッズ」では、マヨネーズの歴史、製造工程、簡単な料理レシピをイラストを使って紹介しています。

Our site targets young homemakers and kids, so it was designed to be simple and fun to navigate. The "Special Recipes" page has an easy-to-operate Recipe Search and several other categories, including a Weekly Recipe and recipes using our mayonnaise and salad dressings. The Kewpie Kids pages contain illustrated presentations on the history of mayonnaise, the manufacturing process, and simple recipes.

Kewpie Homepage
マヨネーズワールド
Mayonnaise World
みんなの食卓を豊かにする、人気者のマヨネーズ。
でもその多彩な素顔については、案外知られていません。
マヨネーズに関する情報あれこれや、ちょっとユニークな調理法など、
展望館にも栄養たっぷりのコーナーです。
New!
ヘルシー！キユーピーマヨネーズ
マヨネーズ情報
マヨネーズ誕生物語
キユーピーマヨネーズの歴史
New!
マヨネーズレシピ
マヨネーズQ&A
全国マヨネーズ協会
オープンキッチン
戻る

Kewpie Homepage
よいこのベビーフード
Baby Food
赤ちゃんの健やかな成長のために
赤ちゃんに食べる喜びを知ってもらうために
ベビーフードについて
商品紹介
ベビーフードができるまで
赤ちゃんの食卓
オンラインショッピング
ベビーフード情報を紹介します。アイコンをクリックしてください。
戻る

Kewpie Homepage
介護食応援団
Care Food
介護食商品紹介
介護食レシピ
介護食情報
介護食Q&A
キユーピーの介護食情報を紹介します。アイコンをクリックしてください。
戻る

Kewpie Homepage
キユーピーチャンネル
Kewpie Channel
CM紹介
テレビ・ラジオ放送ガイド
キユーピーキッズ
ジュークボックス
イベント情報
Kewpie Kids
JUKE BOX
戻る

Kewpie Homepage
キッチンマガジン
Kitchen Magazine
キッチンで役に立つ、なにかと
オイシイ情報が満載！
気軽に読んでいるうちに、
いつのまにかあなたも
"キッチン博士"。
キッチンに立つのが、
もっと楽しくなりますよ。
Table coordinate
Gardening
タマゴのおいしい話
Kewpie press
Vegetables season calendar
Idea
タマゴのおいしい話
Water
戻る

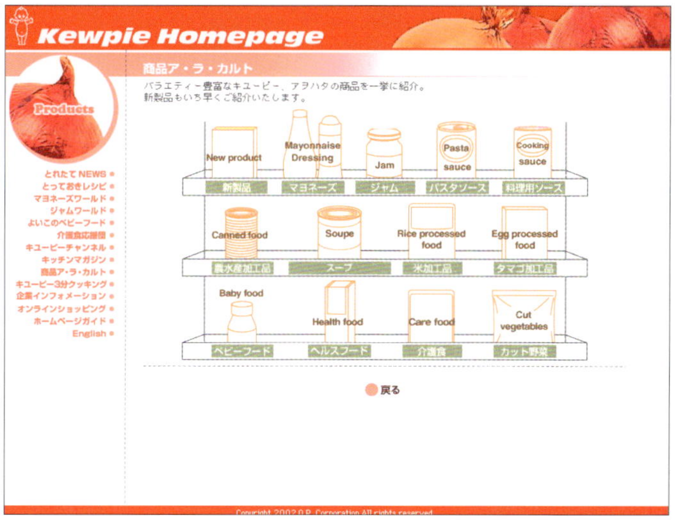

Kewpie Homepage
商品ア・ラ・カルト
Products
バラエティー豊富なキユーピー、アヲハタの商品を一挙に紹介。
新製品もいち早くご紹介いたします。
New product
Mayonnaise Dressing
Jam
Pasta sauce
Cooking sauce
新製品
マヨネーズ
ジャム
パスタソース
料理用ソース
Canned food
Soupe
Rice processed food
Egg processed food
農水産加工品
スープ
米加工品
タマゴ加工品
Baby food
Health food
Care food
Cut vegetables
ベビーフード
ヘルスフード
介護食
カット野菜
戻る

● **S&B**
● **http://www.sbfoods.co.jp/** ● **JAPAN／2002**
CL, SB, DF： エスビー食品（株） S&B FOODS INC. DF： 電通［スーパー・ウェブ（有）］ DENTSU INC. (Super Web,Inc.)

concept 「スパイスとハーブはS&B」、新たにS&Bが発信し始めたメッセージをいかに表現するかをポイントにデザインしました。真中にシンボルロゴを置き、左右に月代わりでスパイスとハーブがレイアウトされます。ナビゲーションもメインとサブを区別し、動線も考えてイアウトしています。

Site design emphasizes S&B's new slogan "S&B, the Spice and Herb Company." The top page layout features a spice and herb of the month pictured to the right and left of the logo at the center of the page. The site is navigated with main and sub menus that incorporate the concept of flow sequence.

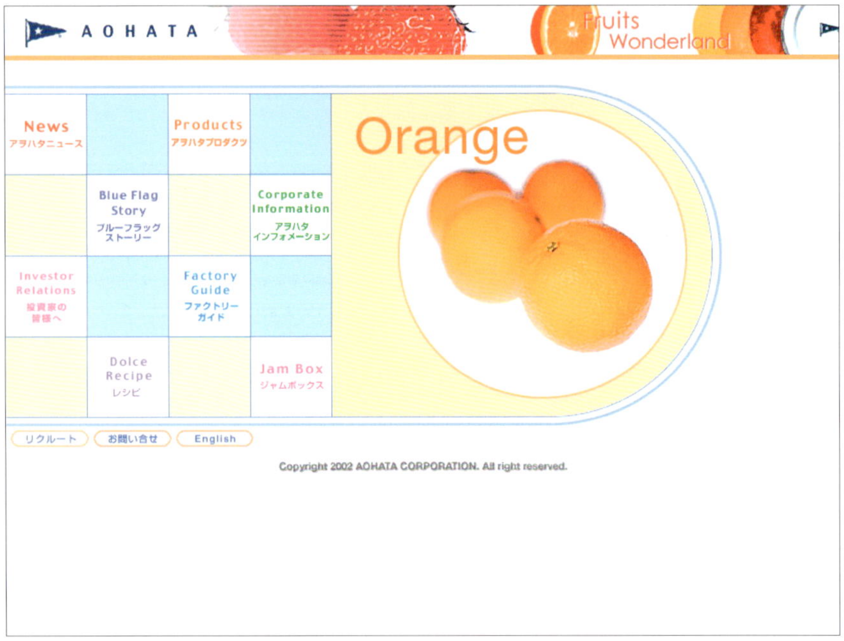

● **アヲハタ（AOHATA）** ● **http://www.aohata.co.jp**
● **JAPAN／2000**
CL, SB： アヲハタ（株） AOHATA CORPORATION DF： （株）朝日広告社 Asahi Advertising Inc.

concept フルーツの彩りや楽しさを表現するようなデザインを意図しました。

The site was designed to accentuate the colorful and flavorful appeal of fruit.

● チェスコ(chesco) ● http://www.chesco.co.jp ● JAPAN／2001

CL, SB：チェスコ(株) CHESCO LTD. DF：(株)エイジア AZIA CO., LTD.

concept 『知って役立つチーズ情報サイト』としてナチュラルチーズの普及啓蒙とナチュラルチーズを気軽に簡単に毎日の食卓に利用していただくためにレシピを中心としたサイト作りをしています。一度見たら忘れられないチーズ色のTOPページも当サイトの売りです。

We created this informational Web site to promote greater familiarity with cheese and encourage people to eat more natural cheeses. The site has lots of recipes that make it easy for people to include natural cheeses in their daily diet. Our "cheese-colored" top page makes a lasting impression on visitors.

● 小岩井乳業(Koiwai Diary Products) ● http://www.koiwaimilk.com
● JAPAN／2001

CL, SB：小岩井乳業(株) KOIWAI DAIRY PRODUCTS CO.,LTD. DF：(株)アサツー ディ・ケイ ASATSU-DK INC.

concept 清潔感と楽しさに満ちた食シーンをイメージいただけるように、全体を明るくやわらかな色調で上品にまとめました。どなたにも親しんでいただきやすいように簡潔明瞭なページ構成にしています。新商品のご紹介やキャンペーンのご案内など、タイムリーな情報をお届けしているほか、関連会社のキリンビール(株)や小岩井農牧(株)にもリンクが可能です。

Light, soft tones create a genteel look that visitors will associate with clean and enjoyable mealtime scenes. We aimed for concise, clear page construction so that everyone who accesses the site will find it comprehensible and enjoyable. The site contains information on new products and campaigns, the latest news, and links to our associates Kirin Brewery Co., Ltd. and Koiwai Noboku Kaisha, Ltd.

● Campbell's ● http://www.campbellsoup.com/ ● USA／2002
CL, SB：Campbell Soup Company.

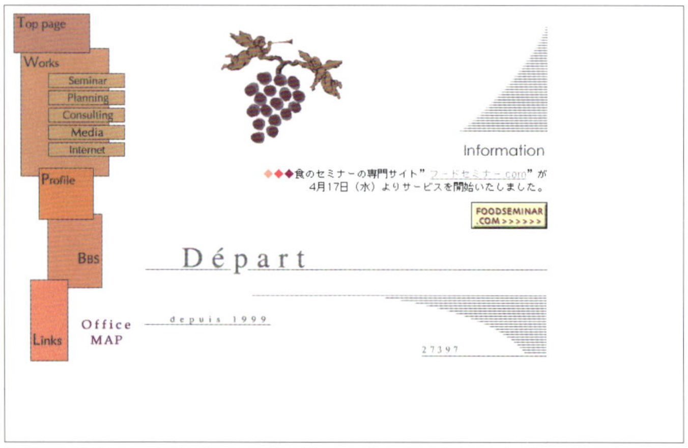

● デパール(depart, inc.) ● http://www.depart-vin.com/ ● JAPAN／2000
CL, SB：(有)デパール depart, inc.

● BARBARA'S BAKERY ● http://www.barbarasbakery.com/
● USA／2002
CL, SB：Barbara's Bakery Inc.

● BOUDIN BAKERY ● http://www.boudinbakery.com/ ● USA／2001
CL, SB：BOUDIN BAKERY

● ドミノ・ピザ（Domino's Pizza）
● http://www.dominos-pizza.co.jp/
● JAPAN／2001
CL, SB：（株）ヒガ・インダストリーズ ドミノ・ピザ本部 Higa Industries Domino's Pizza
DF：（株）EXIT EXIT Co., Ltd.

concept ドミノ・ピザの最新ニュースやキャンペーン情報を紹介。メニュー紹介や最寄りのお店を検索できる情報満載のメルマガ「DoPiDo！」（ドゥー・ピー・ドゥ）を発行しています。

Our site has the latest Domino's Pizza news, campaign information, and a complete menu. Users can search for Domino's Pizza stores in the area where they live. We also issue a mail magazine, DoPiDo!, that's packed with information.

● the old spaghetti factory ● http://www.osf.com/
● USA／2002
CL, SB：OSF International, Inc.

● デルモンテ（Del Monte） ● http://www.kikkoman.co.jp/delmonte/
● JAPAN／2002
CL, SB：キッコーマン（株） Kikkoman Corporation.

- リプトン（Lipton）
- http://http://www.lipton.co.jp
- JAPAN／2002

CL, SB：日本リーバ（株） Nippon Lever K.K.
DF：J.ウォルター・トンプソン・ジャパン（株） J. Walter Thompson Japan(=JWT)

concept リプトンの持つ明るくてオシャレな世界観を、おもちゃ箱のような背景とヒップで楽しげなキャラクター達で表現しました。また「クリックする楽しさ」を思い出してもらう新鮮なクリックアクションやギミックの数々や、「ページ」という概念を廃したインターフェイスデザインで新しいウェブインターフェイスを志向しています。

Lipton's bright, stylish world is portrayed with hip, cheerful cartoon characters and a background that looks as if it was made out of the contents of a child's toy box. We've made it fun to navigate the site by combining clicking with animation and a variety of fun gimmicks, and created a new type of interface design that shifts away from the traditional "page" concept.

● **UCC**
● **http://www.ucc.co.jp/jin/**
● **JAPAN／2001**
CL, SB：UCC上島珈琲（株） UCC Ueshima Coffee Co., Ltd.
DF：（株）デジタル・マジック・ラボ Digital Magic Labs

concept コーヒーを愛してやまない人、これからコーヒーを知りたいと思っている人。・・・コーヒーに興味のある人は誰でもみんな“珈琲人”。そんな珈琲人たちに心をこめてお届けするコーナーです。

People who love coffee, people who want to learn more about coffee-anyone interested in coffee is a "Coffee Person!" We put our heart into creating a Web site that all Coffee People can enjoy.

● **Folgers** ● **http://www.folgers.com/** ● **USA／2002**
CL, SB：Folgers

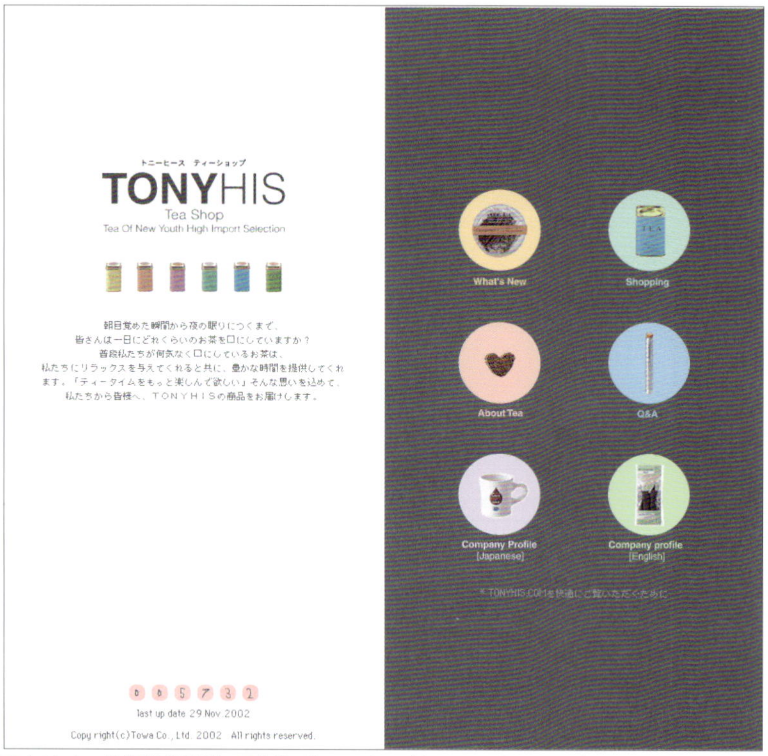

● トニーヒース(TONYHIS) ● http://www.tonyhis.com ● JAPAN／2002

CL：(株)陶和 Towa Co., Ltd. SB：トニーヒース TONYHIS DF：(株)陶和 Towa Co., Ltd. (Design Director) / (株)mazzic mazzic

concept 20～30代の女性がメインターゲット。特に、ブライダル向け商品やギフトを探している客層開拓に力を入れています。ダークグレイのメインカラーと、女性を意識した淡い色との組み合わせで、女性が入りやすいサイトイメージを作っています。一つ一つの商品に対する丁寧な説明やお茶に関するコラム等を盛り込み、より紅茶を身近に感じてもらえたら、と思っています。

The site targets young women in the 20s and 30s age group. In particular, we want to expand our customer base by presenting tea and tea accessories as an attractive bridal gift option. The top page is predominately dark gray combined with pastel colors that appeal to women, in order to draw them into the site. To help users become more familiar with tea, we have included a detailed description of each of our products and articles on topics related to tea.

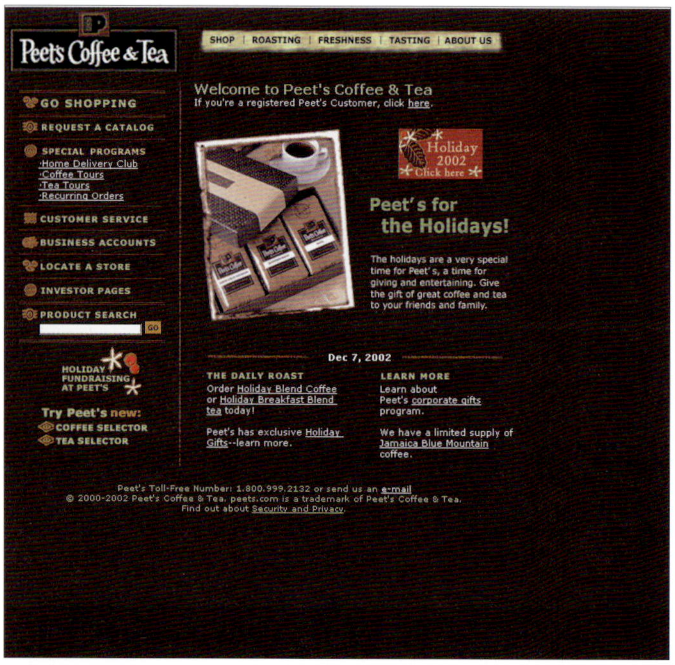

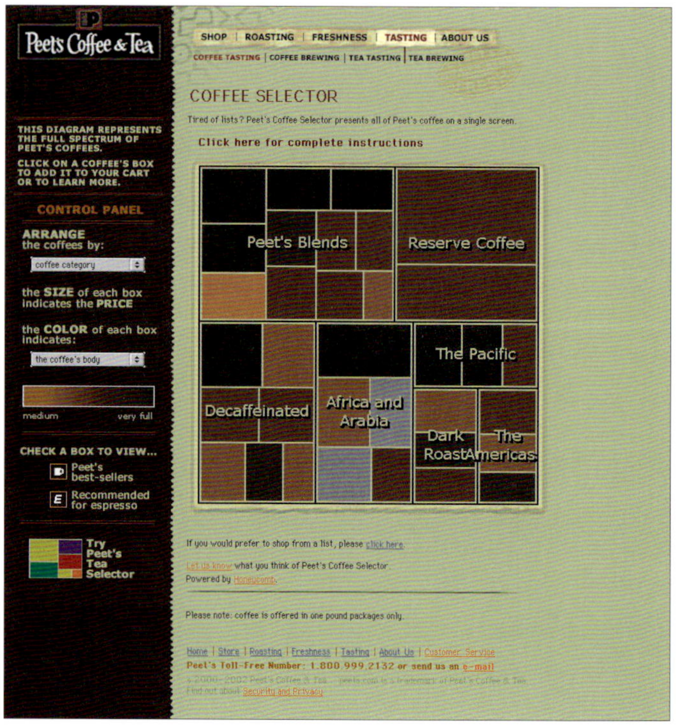

● Peet's Coffee & Tea ● http://www.peets.com/ ● USA／2002

CL, SB：Peet's Coffee & Tea

● コカ・コーラ(Coca-Cola) ● http://www.cocacola.co.jp/
● JAPAN／2001
CL, SB：日本コカ・コーラ(株) Coca-Cola Japan Company Limited DF：電通 DENTSU INC.

concept ブランド各サイト、コーポレイト関連ページのポータルとしての位置付け。キャンペーン情報はアニメーションが告知し、詳細ページへ誘導します。

This is a portal site to our brand Web sites and corporate-related Web pages. Animation is used to announce campaigns and draw visitors to pages with detailed campaign information.

● Sprite ● http://sprite.rocketcash.com/ ● USA／2002
CL, SB：The Coca-Cola Company.

● evian ● http://www.evian.com/ ● FRANCE／2002
CL, SB：EVIAN

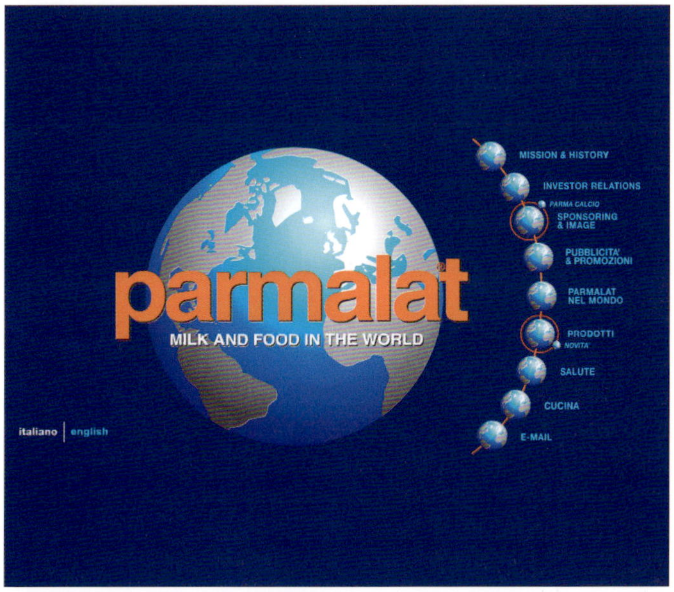

● Parmalat ● http://www.parmalat.com/ ● ITALY／2002
CL, SB：Parmalat

● Asahi ● http://www.asahibeer.co.jp/
● JAPAN／2002
CL, SB：アサヒビール（株） ASAHI BREWERIES, LTD. DF：（株）ガーラ GALA Incorporated

concept ビールという商品のイメージや弊社の訴求したいイメージである、「元気」「挑戦」「楽しさ」といった要素を動的なフラッシュを利用して表現しました。またHP全体のナビゲーションとして、ユーザーの利用動機を意識して、「商品」「会社」「おすすめ」「What's new」といった構成にしています。

The site uses Flash to dynamically communicate an image for our beer products and the company of health and vigor, challenge, and enjoyment. The site is organized into four sections-Products, Corporate Information, What's New, Fun Information-for easy navigation and to accommodate the varied interests and purposes of visitors.

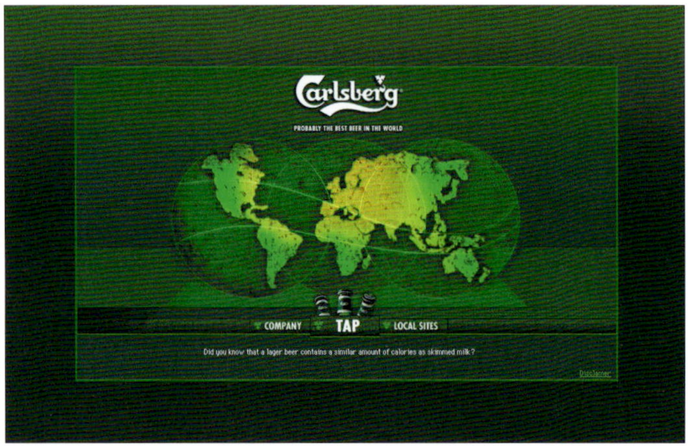

● Carlsberg ● http://www.carlsberg.com/ ● DENMARK／2002
CL, SB：Carlsberg A/S

● Mountain Dew ● http://www.mountaindew.com/ ● USA／2002
CL, SB：PepsiCo, Inc.

● Dr Pepper ● http://www.drpepper.com/ ● USA／2002
CL, SB：Dr Pepper / Seven Up, Inc.

● KIRIN ● http://www.beerdaigaku.com/
● JAPAN／2001
CL, SB：キリンビール（株） Kirin Brewery Co., LTD. DF：（株）葵プロモーション Aoi Advertising Promotion Inc.

concept キリンビール大学は、インターネット上に開校した、ビール文化を広く多くの方々に知ってもらうことを目的とする、情報提供型のアクセスフリーサイトです。サイトのビジターに楽しく学んでもらうために、大学というモチーフを設定し、講義という形でビールに関する様々な情報を発信しています。また、サッカー部といったクラブ活動や、実体験型の課外授業なども年に数回開催しています。

Kirin Beer University is an informational site we opened to spread the beer culture to as many people as possible. Anyone can enter (access) the university for free. We chose the "university" motif to make it fun for visitors to learn about beer. Information is presented in the form of university "lectures." Users can also have fun participating in a virtual soccer club. The site includes a page for students to write about actual experiences at Kirin Beer University seminars held periodically during the year.

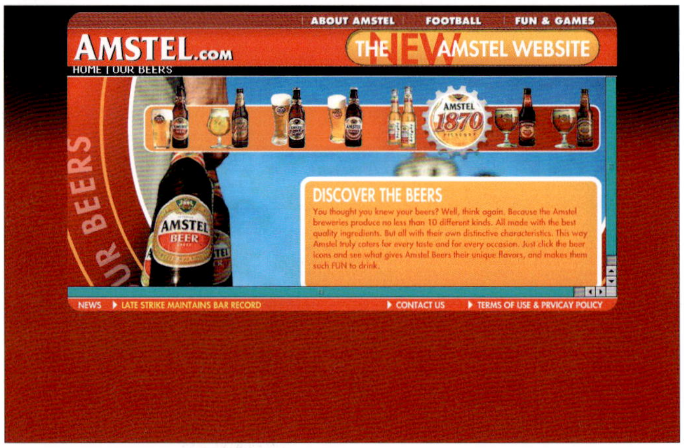

● AMSTEL ● http://www.amstel.com/ ● THE NETHERLANDS／2002
CL, SB：Amstel

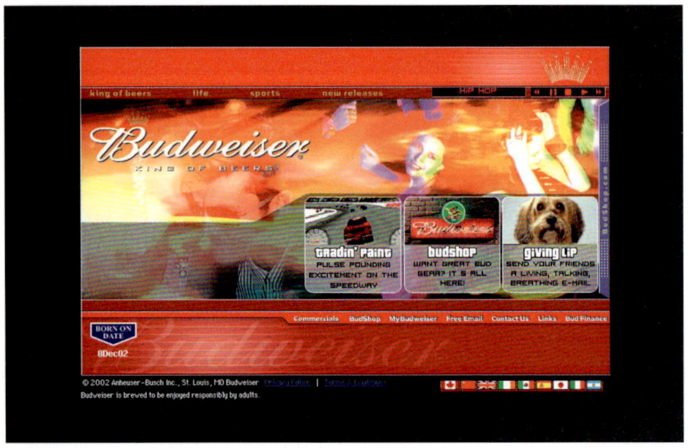

● Budweiser ● http://www.budweiser.com/ ● USA／2002
CL, SB：Anheuser-Busch Companies, Inc.

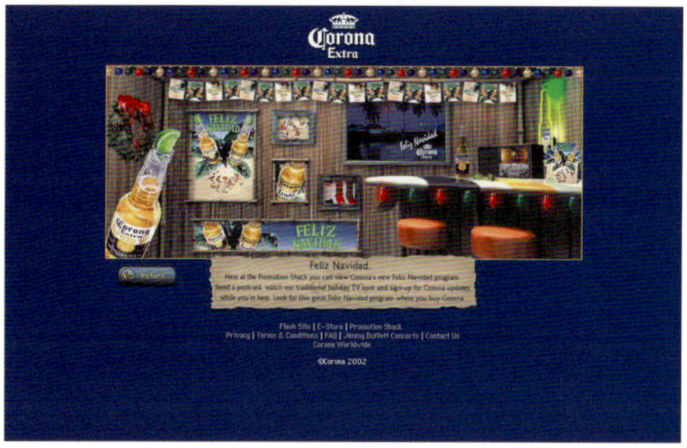

● Corona Extra ● http://www.corona.com/ ● MEXICO／2002
CL, SB：Corona

● TUBORG ● http://www.tuborg.com/ ● DENMARK／2002
CL, SB：Carlsberg A/S

● BAWLS GUARANA ● http://www.bawls.com/ ● USA／2002
CL, SB：BAWLS Guarana

● SAMUEL ADAMS ● http://www.samadams.com/ ● USA／2002
CL, SB：Boston Beer Company

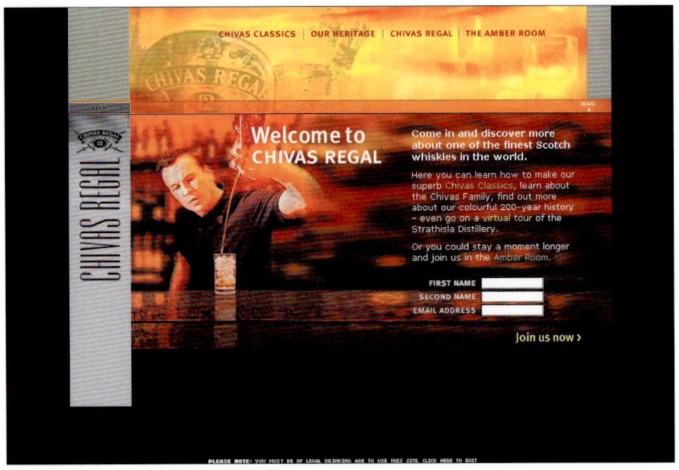

● **CHIVAS REGAL** ● **http://www.chivas.com/** ● **UK／2002**
CL, SB：Chivas Brothers

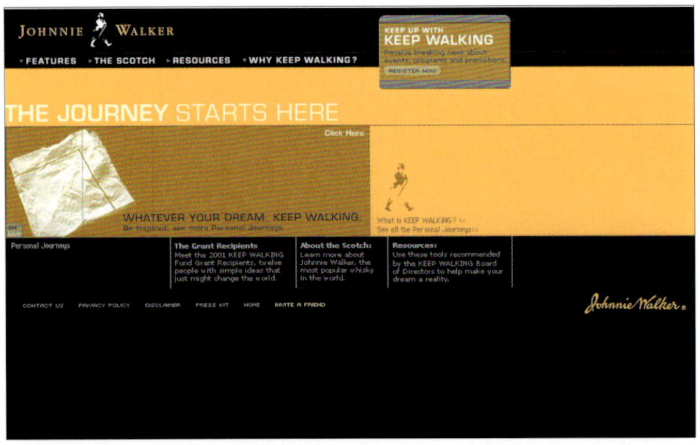

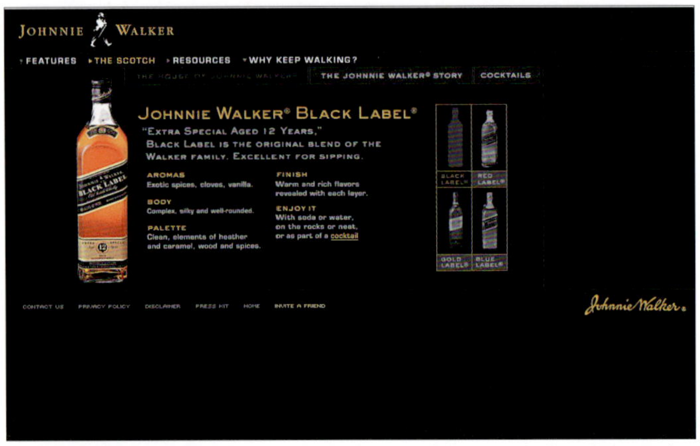

● **JOHNNIE WALKER** ● **http://www.keepwalking.com/** ● **SCOTLAND／2002**
CL, SB：JOHNNIE WALKER

● **WILD TURKEY**
● **http://www.wildturkeybourbon.com/**
● **USA／2000**
CL： STC Corporation SB： Pernod Ricard USA

concept バーボンウィスキーをこよなく愛する人々への情報発信と交流を目的として開設しました。ワイルドターキーの歴史、遺産、主義が紹介されているほか、Turkey Gearをオンラインで購入することもできます。

The Wild Turkey Web site was created and designed to promote interaction and involvement with Bourbon Whiskey consumers and to inform consumers about the history, heritage, and tradition of Wild Turkey. The site is e-commerce capable, so customers can purchase Wild Turkey merchandise.

● **Furuta** ● **http://www.furuta.co.jp/** ● **JAPAN／2001**
CL, SB： フルタ製菓（株） Furuta Confectionery Co.,Ltd. DF： （株）コンテンツジャパン CONTENTS JAPAN

concept 弊社のスローガン「おいしさと夢を」を赤（食欲をそそる色）を基調とし、デジタルではなくアナログ（手書きでレタリングしてインターネットの画像として使用）で表現しています。

The Internet is a digital environment, but our slogan "Deliciousness and Dreams" is analog-the handwritten image was scanned in and placed against a red background. The site is done primarily in red, because it is a color that stimulates the appetite.

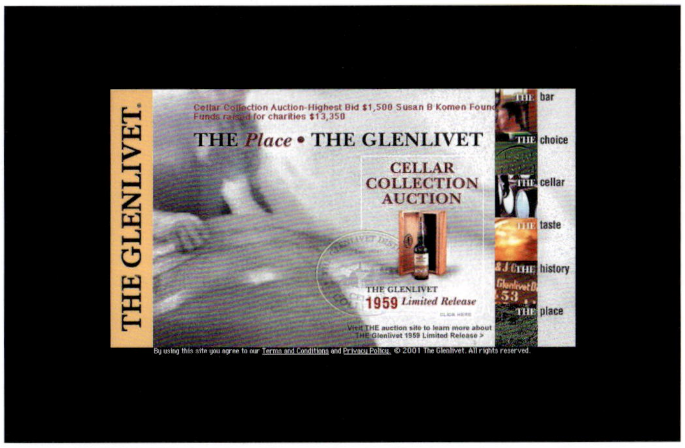

● **THE GLENLIVET** ● **http://www.theglenlivet.com/** ● **UK／2002**
CL, SB： Civas Brothers

● **清月堂本店（Seigetsudo-Honten）** ● **http://www.seigetsudo-honten.co.jp/**
● **JAPAN／2002**
CL, SB： 清月堂本店 Seigetsudo-Honten

● モロゾフ(Morozoff) ● http://www.morozoff.co.jp/ ● JAPAN／1999
CL, SB：モロゾフ（株） Morozoff Co., Ltd.
DF：（株）パワー・インタラクティブ POWER INTERACTIVE CORP.

concept 弊社の企業イメージをわかりやすくご理解いただけるようトップページ（メニューページ）をビジュアル豊かなページとしています。また、企業テーマである「ロマンのあるスイート」をサイト上で表現できるよう単なる商品紹介にとどまらず、スイートのある生活提案や、原材料、モノづくりへのこだわりが伝わるコンテンツ作りを目指しています。

The top page (menu page) is made into the visual rich page so that you can understand the corporate image of our company intelligibly. Moreover, it does not remain in a mere goods introduction so that "the existing romantic sweet" who is a company theme can be expressed on a site, but it aims at making the contents from which the prejudice by the existing sweet life proposal, and raw material and the production of monochrome is transmitted.

● 不二家(FUJIYA) ● http://www.fujiya-peko.co.jp/
● JAPAN／2002
CL, SB：(株)不二家 FUJIYA CO.,LTD. DF：モーフグラフィックス morph graphics

concept 2002年秋にリニューアルした“不二家ホームページ”。“かっこよさ”と“可愛らしさ”を兼ね備えたサイトデザインが特徴です。今回のリニューアルでは初心に返り、本業であるお菓子やケーキといった商品情報のユーザビリティ(検索の利便性やページ容量の軽さ)に特にこだわりました。言い換えるなら「不二家総合図鑑」、本業と連動した新鮮で楽しい情報を発信していきます。

We renewed the Fujiya Web site in the fall of 2002. The new design blends "stylish" and "cute" elements. We've gone back to basics to construct a site that concentrates on providing information on Fujiya's core productsŃcandy, cookies, and cake. We also emphasized usability; the site is easy to search and pages load quickly. You might describe the new site as a "Fujiya Picture Book" full of new and fun information on our core business.

HOME

Furniture & Interior
［家具＆インテリア］

Electrical Products
［電化製品］

Kitchen appliances
［キッチン器具］

Lifestyle Products & Sundries
［生活用品＆雑貨］

Cars, Motorcycles & Bicycles
［自動車、バイク＆自転車］

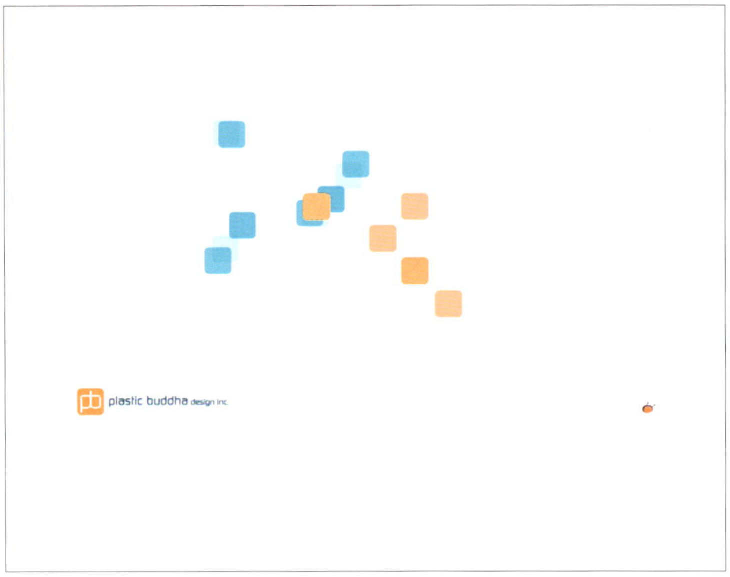

● plastic buddha design inc.
● http://www.plasticbuddha.com ● CANADA／1999-2000
CL：Plastic Buddha Industrial Design Inc. SB：spacecadet design inc. DF：Chris Clarke

concept

plastic buddhaは人々の予測を大胆に裏切ることで世界的な評価を確立してきた産業デザイン企業です。そのデザインは非凡どころか時に常軌を逸しているようにも見えますが、美的感覚と機能性が見事に調和しています。Webサイトの構築でも同じ評価を得ることを目指し、ロード中のコンテンツを意図的に見せる斬新なスタイルを採用しました。このとき共に表示されるコラージュはWebサイトのあちこちに表示され、トレードマークの役割を果たしています。メインページは水槽の中をフワフワと泳いでいるかのようで、コンテンツを見たい気持ちを駆り立てながらも、どこをクリックすればよいのかわかりません。しかし点滅する四角形にカーソルを重ねれば、そのアイコンが訪問者との対話への入口であることがすぐにわかります。plastic buddhaのデザインオブジェクトは、一貫したスタイルの製品化というより、一貫した概念・思考や問題の解決方法を製品化したものと言えます。したがって、セクションごとにユニークなデザインが施されているにもかかわらず、全体的なデザインの雰囲気を逸脱していません。「Ugli」セクションには清潔感のある美しさがあり、「Porno」セクションで点滅する画像は劣化した古い映像をプロジェクターで投影しているかのようです。「Press」セクションでは、引用句が画面を滑るように表示され、訪問者の関心を引きつけます。このWebサイトを閲覧するのは販売店、メディア関係者、同業他社たちであるため、企業としての概念を明確に打ち出すよう、工夫が凝らされています。

The Web site of plastic buddha, an internationally acclaimed industrial design company that has gained notoriety because its products intentionally challenge people's expectations. Although the design objects defy convention and sometimes seem impossible, they are actually quite functional and beautiful. We wanted the Web site to have the same sort of feeling. The loading screen is atypical because it doesn't try to hide what it is loading; instead it creates a collage of imagery that the user sees throughout the Web site. The main page interface is designed to be hypnotic, like fish in a fish tank, and yet compelling enough to encourage exploration. This works so that users are initially disoriented and unsure of what to do; but once they roll the mouse over the flashing squares the navigation method becomes immediately clear, suggesting that the square icon is the primary means of interaction. Design objects from plastic buddha are the product of consistent conceptual thinking and problem solving rather than consistent styling. Each section was given a unique look, but fits within the overall design context. The Ugli section is clean and beautiful. The Porno section contains flickering imagery that is reminiscent of seedy movies viewed with a projector. The Press section slides quotes across the screen to maintain user interest. This is a site that is viewed by retailers, journalists, and fellow industrial designers, so care was taken to communicate company concepts clearly.

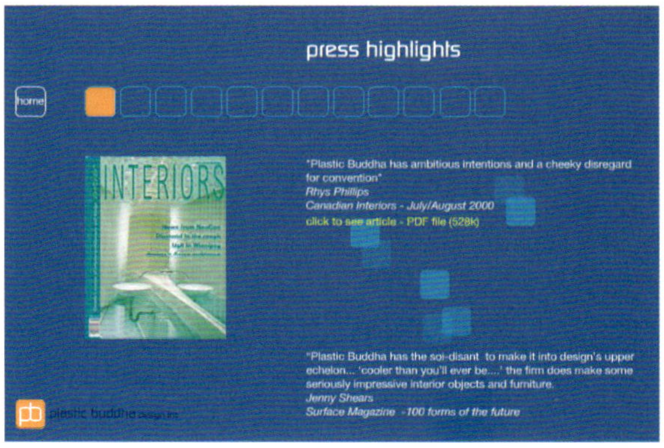

SCANDINAVIAFORM
http://www.scandinaviaform.com
SWEDEN／2002
CL. SB：SCANDINAVIAFORM Ltd　DF：Maxjenny Superstudio

concept

製品の一例を見ただけで好奇心をそそられるWebサイト。どのページを見てもテンポよく詳細が表示されます。フレームにランダムに配置された四角形のパターンに、ポップな色使いと繊細なフォルムが配置されています。Scandinavianformのビジョンをまず理解してもらうために、「Shop Online」セクションをクリックするまで製品が表示されません。テキストの赤い色はフレームの色と見事に調和しており、クリアで簡潔なナビゲーションは、訪問するすべての人がわかりやすいと感じるようデザインされています。スムーズ、クール、そしてフレッシュなデザインは、新しい北欧デザイナーの手による傑作を展示するにはベストの方法です。北欧デザインの新世代を担うデザイナーの作品を紹介し、これまで北欧デザインに縁のなかった訪問者の購買意欲を駆り立てることを目的に作成されました。

The idea is to spark the viewer's curiosity with just a glimpse of a few chosen products—a detail here and a detail there shown with good, animated rhythm. Strong colors and sensitive forms shift in square frames in a random pattern. Visitors cannot see the products in their entirety until they enter the Shop Online section. Hopefully users will read about Scandinaviaform's vision first. The text is red to harmonize with the shifting frame colors. Clear, straightforward navigation was designed to be comprehensible to everyone who visits the site. The smooth, cool, and fresh look is compatible with the site's function as a display vehicle for the best products from new Scandinavian designers. The site's purpose is to introduce groundbreaking Scandinavian design straight from a new generation of Scandinavian designers, and bring the Scandinaviaform shopping experience to people to whom it would not otherwise be available.

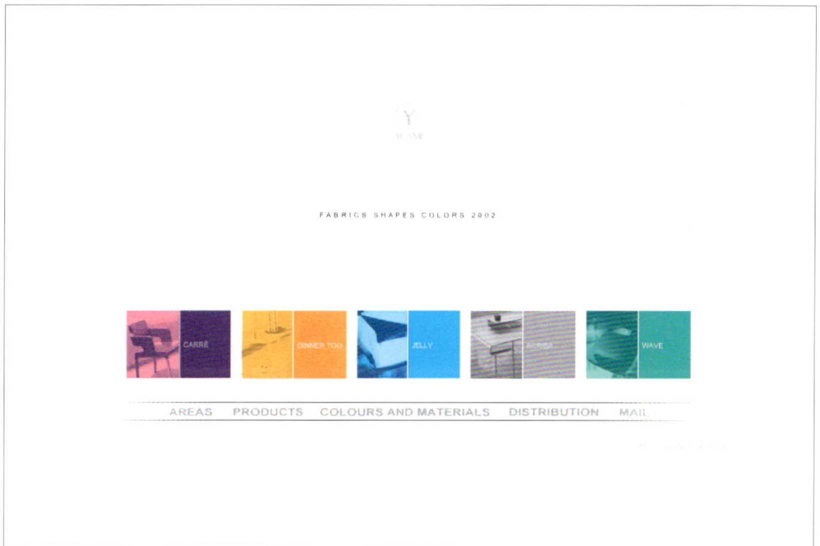

YCAMI　**http://www.ycami.com**　**ITALY／2002**
CL：YCAMI　SB：YCAMI S.p.A.　DF：IMPULSE

LINEA　**http://www.linea-inc.com/**　**USA／2002**
CL. SB：LINEA

● OFFI ● http://www.offi.com
● USA／2002
CL：OFFI & Company SB：Cottage Industries DF：Eric Pfeiffer

concept トップページを一目見ればOFFIの雰囲気が伝わるWebサイトです。4つのカテゴリにカーソルを重ねてみてください。そのカテゴリのサムネイルが表示されます。目的のサムネイルを見つけたら、カーソルを重ねるとアイテムリストが表示され、製品名にカーソルを重ねると製品の画像が大きく表示されます。ここでクリックすれば、別ウィンドウで詳しい説明が表示され、カートにも追加することができます。訪問者がコンテンツに集中できるように、色使いやフォントを工夫し、空間を広く設けました。

We want Web site visitors to get the flavor of OFFI with the first menu. The site is easily navigated by mousing successively over the four main categories, sub-category thumbnails, and names in lists of product items. Users click product images to pop up windows with more detailed information or to add products to the shopping cart. Colors, fonts, and white space are all designed to focus site users on OFFI.

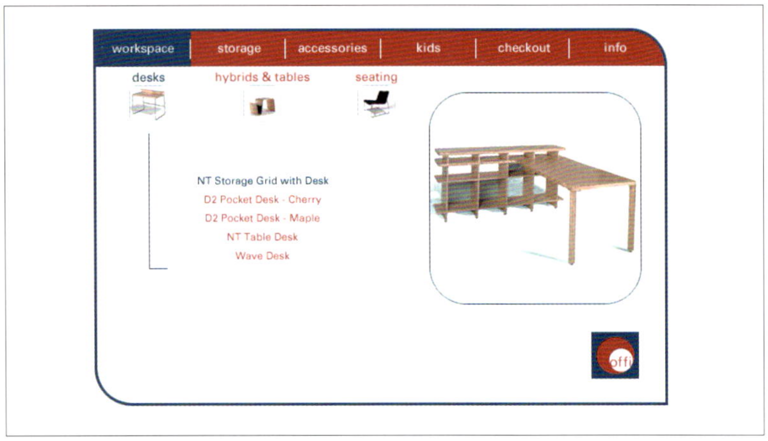

● European Furniture Importers ● http://www.eurofurniture.com/ ● USA／2002
CL, SB：European Furniture Importers

● **Emeco** ● **http://www.emeco.net** ● **USA／2001**

CL, SB：Emeco DF：Denis Tangen / Barbara Folk / Billie Howe / Halie Schmidt

concept あなた自身を変えたくありませんか？　待っていても変われないし、成長できません。自分自身でつくりだし、活用するのです。変わるために必要なもの。それは経験を怖れないオープンな心、そして変わりたいという意思です。

Allow events to change you. You have to be willing to grow. Growth is different from something that happens to you because you are the one who produces growth and lives it. The prerequisite for growth is a willingness to experience events and be changed by them.

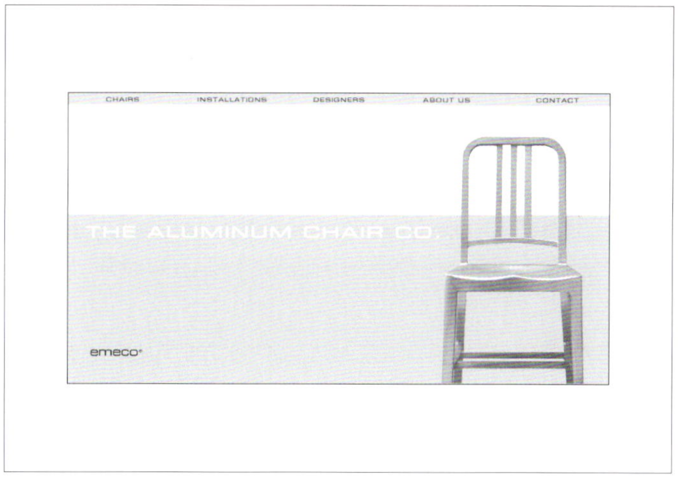

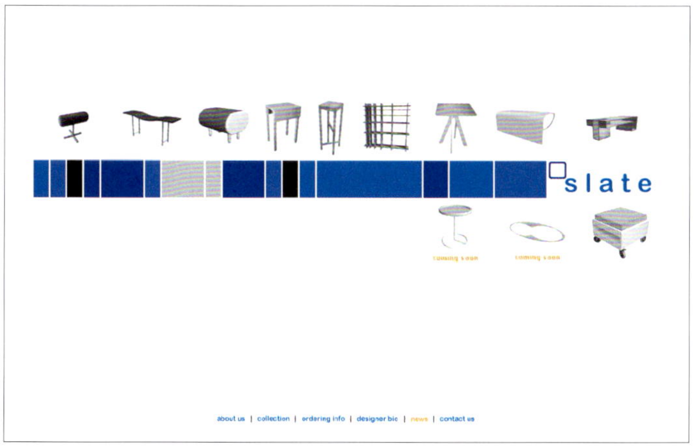

● **slate** ● **http://www.slatedesign.com** ● **USA／1999-2002**

CL, SB：slate DF：Mark Daniel

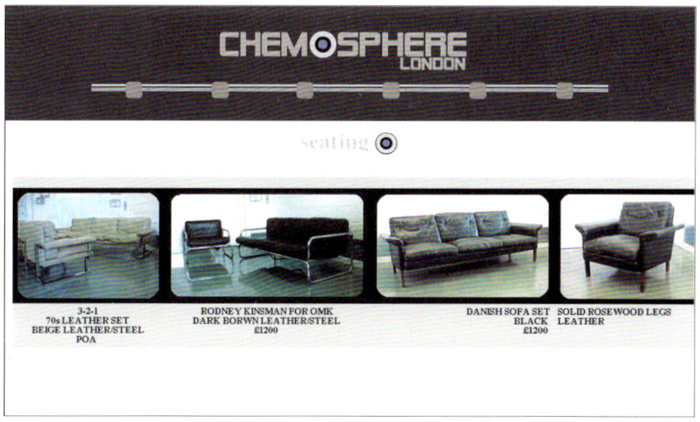

● **CHEMOSPHERE LONDON** ● **http://www.chemosphere.co.uk/**
● **UK／2002**

CL, SB：Chemosphere London

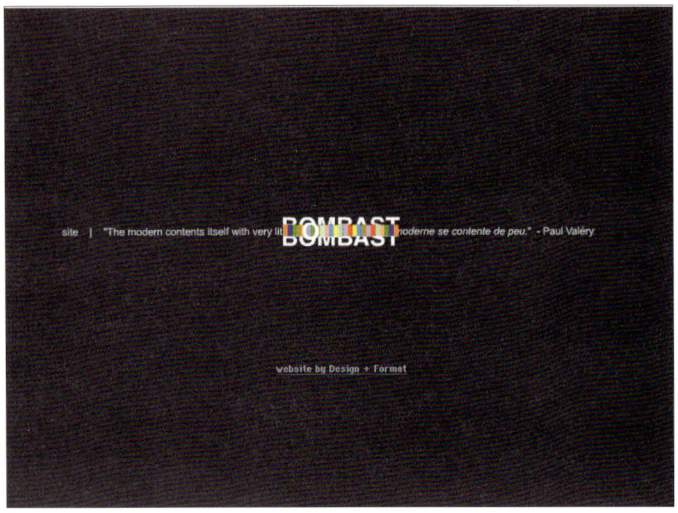

● BOMBAST ● http://www.bombastfurniture.com ● CANADA／2000

CL, SB：BOMBAST FURNITURE LTD.　DF：Lam Wong / Design Format

concept

控えめで古典的、落ち着いていて、表情豊か。さらに家具への敬愛があらわれたWebデザインを目指しました。ここには仕掛けもサウンドも不要です。Lam Wongは、伝統的なグラフィックデザインを学ぶと同時に、新しいメディアにも精通しているWebデザイナーです。彼のグラフィックに対する熱意のおかげで、新旧のギャップが埋められるサイトになったと確信しています。

We wanted a restrained, classically poised, quietly expressive design for the Web site that would complement the furniture—no gimmicks, no bells, and no whistles. Lam Wong is a designer trained in traditional graphic design and thoroughly familiar with new media who brings a dual focus to Web design. His less-than-anxious relationship to the history of graphic presentation has generated a site that we believe successfully negotiates the gap between old and new mediums.

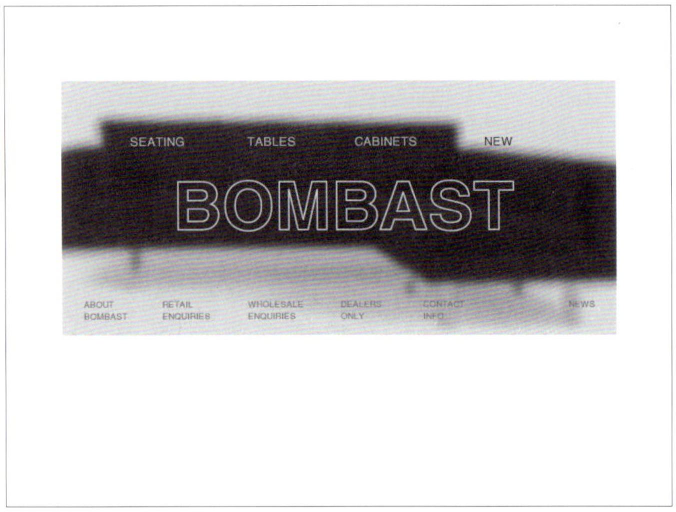

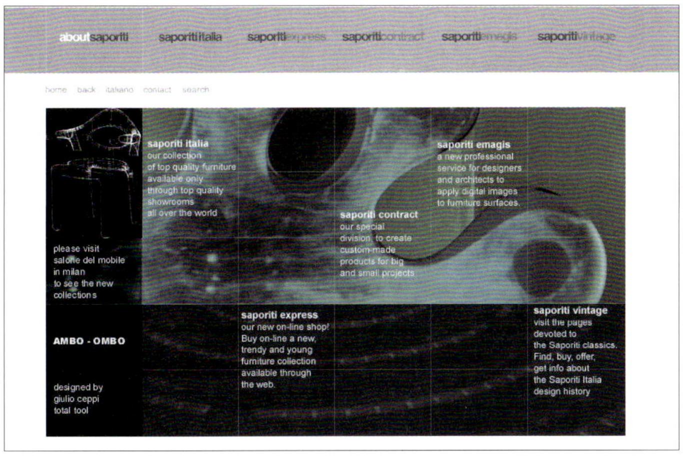

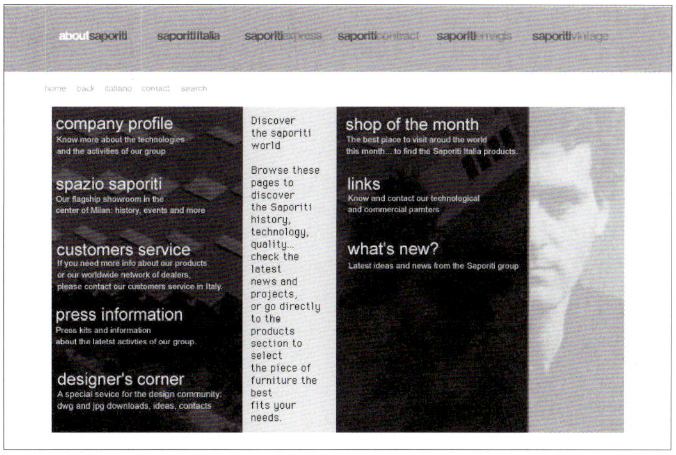

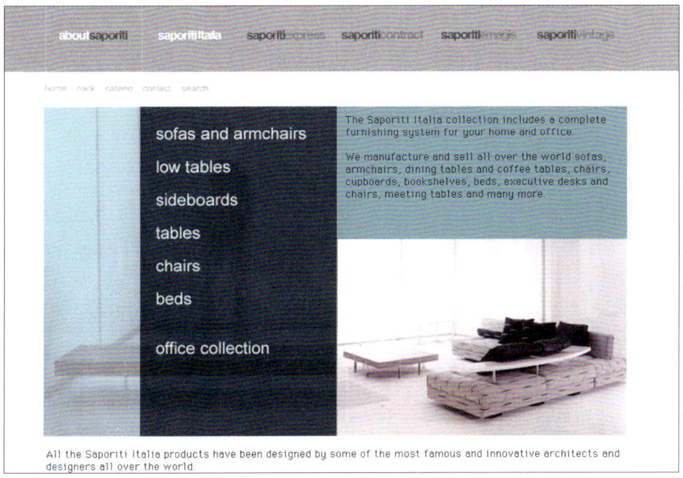

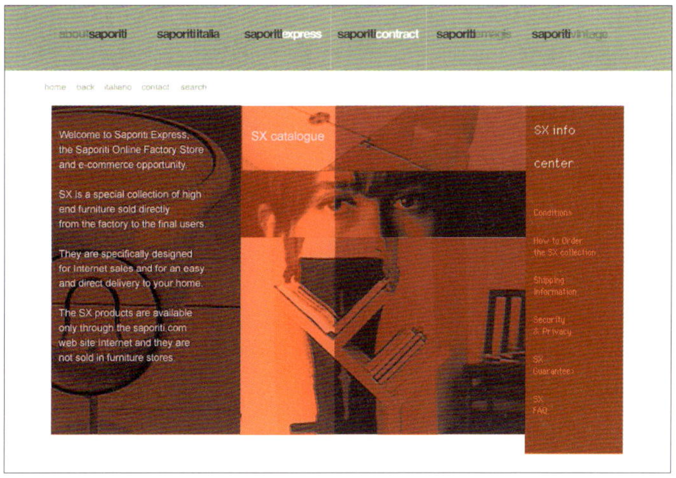

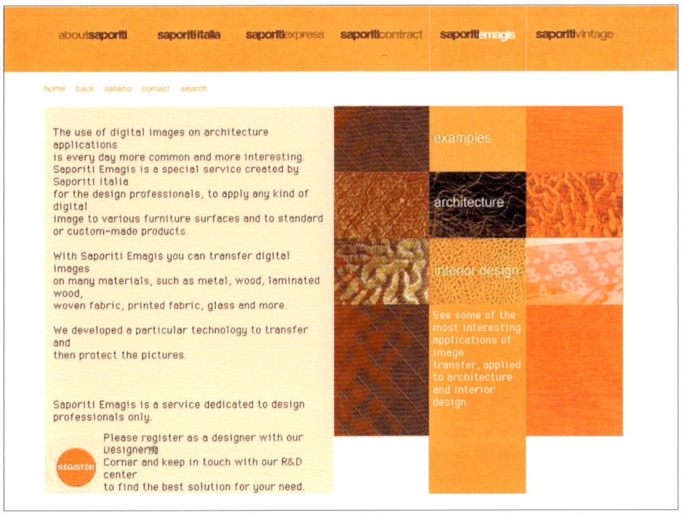

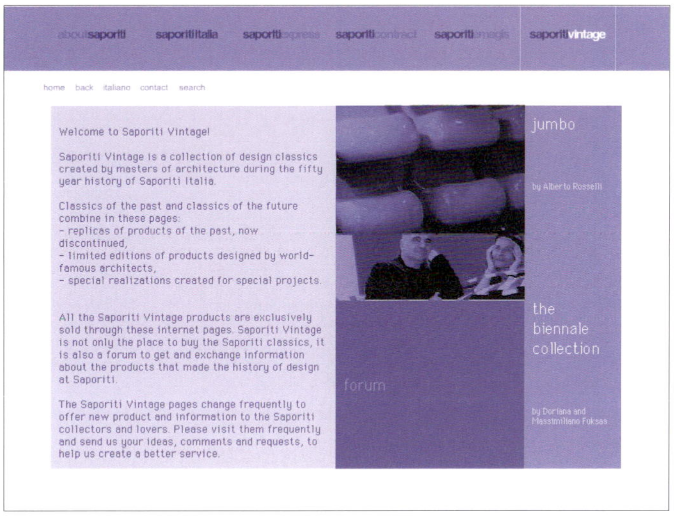

● **saporiti italia** ● **http://www.saporiti.com/** ● **ITALY／2002**
CL, SB : Saporiti italia

● **Eric Janssen Design** ● **http://www.ericjanssendesign.com**
● **USA／2000**
CL, SB : Eric Janssen Design, LLC　DF : Eric Janssen

concept シンプルで拡張可能なツールを目指してデザインしたWebサイトです。製品の一覧と会社概要を掲載しています。全体としてHTMLで作成されているため、初心者でも簡単にコンテンツを修正することができ、小さな修正のたびに財布の中身を気にする必要がなくなります。Webサイトに必要なのは、情報にすばやくアクセスできるシンプルかつコンパクトな構成です。情報がロードされるまで延々と待たなければならなかったり、必要な情報にたどりつくまでに何度もクリックしなければならないようなWebサイトにすべきではないのです。さらにWebはグローバルなネットワークであるため、発信する情報は最小限に抑えるべきだと考えます。あるいは、効率的で説明的な役割を果たすイメージを使用する努力が必要です。そうすれば、日本語が読めなくても大好きな日本語のWebサイトを理解できるのですから。デザインはそれ自体が寡黙な言語なのです。

Our Web site was designed to be a simple, expandable tool that showcases our products and provides information about the company. It is constructed entirely in html for easy modification by even a novice like myself. I hate the idea of having to pay someone for every little change! I believe Web sites should be simple and compact, so information can be accessed quickly and easily. Users shouldn't have to wait an eternity for information to load or navigate through a million screens to find what they are looking for. And since the Web is a global network, written information should be kept to a minimum—or at least supported by effective, self-explanatory visual images. That way you don't have to be able to read Japanese to understand a great Japanese Web site. Design is, after all, its own silent language.

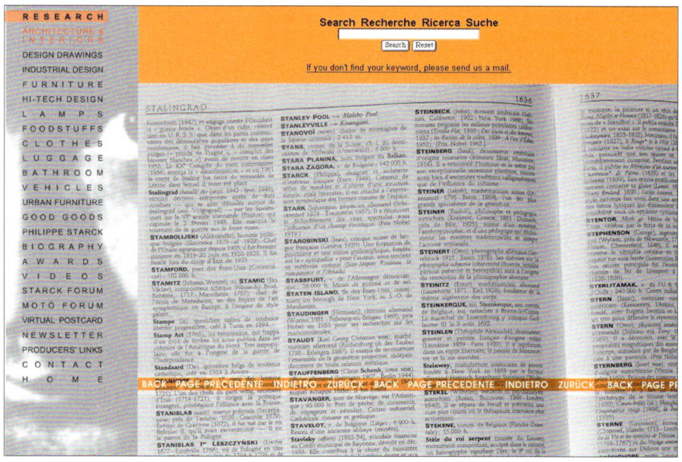

● **Philippe Starck** ● **http://www.philippe-starck.com** ● **GERMANY／2002**
CL : Agency Philippe Starck　SB : Web Workstyle e.K.　DF : Markus Myrach / Antonella Tolotti / Glauco Gatelli

concept 1人のファンから恋する友人へ

From a fan for a friend in love.

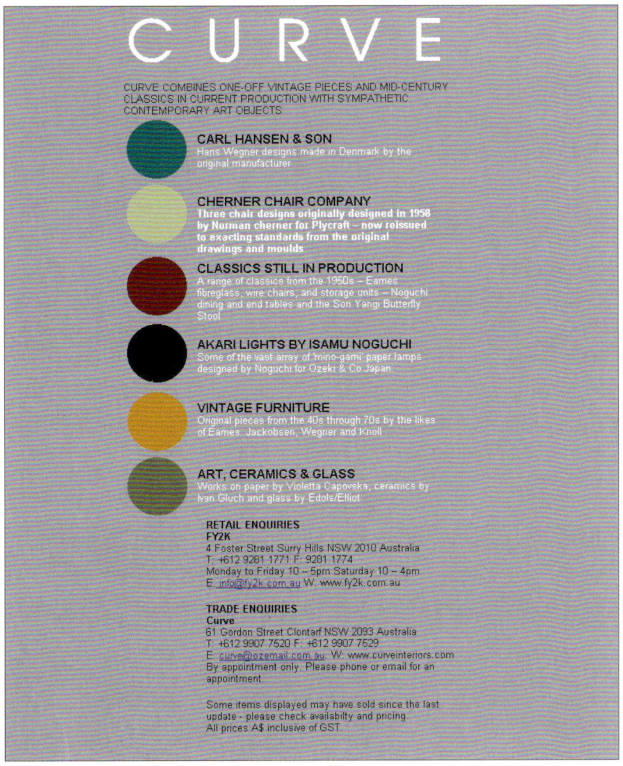

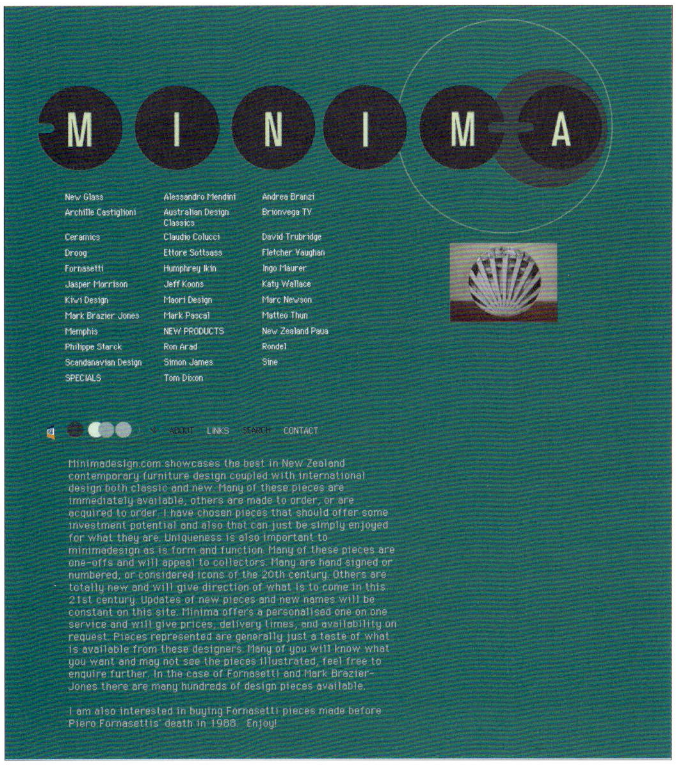

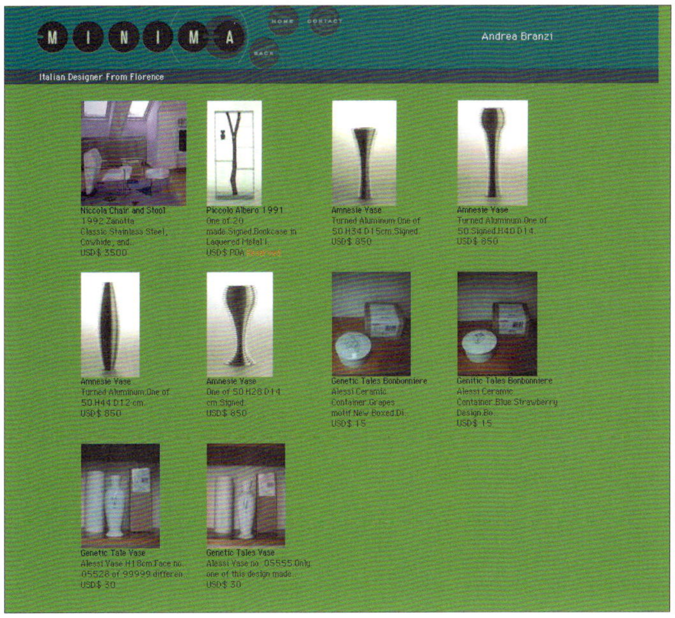

● **CURVE** ● **http://www.curveinteriors.com/** ● **AUSTRALIA／2002**
CL, SB：CURVE

● **Minima Design** ● **http://www.minimadesign.com**
● **NEW ZEALAND／1999**
CL：Minima design SB：www.minimadesign.com DF：Q Cee

concept ニュージーランドをはじめとする各国のデザインを展示。コンテンツを最小限に抑えることをコンセプトとしています。製品はデザイナー別に分類され、美しいビジュアル形式で表示されます。Eコマース機能とGPRS携帯電話技術を採用しているため、お問い合わせに即座に回答することができます。

Minima Design showcases New Zealand and international designs. A minimalist approach was applied to the site design. The site is easy to use and provides clients with the information they want in an aesthetically pleasing visual form. Products are grouped by designer. Minima Design has e-commerce capability, and GPRS cell phone technology enables us to reply instantly to inquiries.

● INTERIOR INTERNET ● http://www.interiorinternet.com/ ● UK／2002
CL, SB：INTERIORINTERNET

● OUT.DeSIGN ● http://www.outdesign.com/ ● JAPAN／2000

CL：（株）アウトデザイン 黒川 勉 OUT.DeSIGN Tsutomu Kurokawa SB：（株）ブルーマーク Bluemark Inc. DF：（株）ブルーマーク ／ 小林シンヤ（SUP） Bluemark Inc. / Shinya Kobayashi (SUP)

concept 数多くのポートフォリオを直感的に見せることが目標。Flashによる横方向の年表形式というアイデアを選択。結果、黒川勉のキャリアを感覚的に把握できるツールとなり、特に海外向けのポートフォリオとして効果的に機能しています。

The idea was to construct a site that presents the artist's large portfolio so that viewers get an immediate sense of the work. The horizontal timeline, created with Flash, is a tool that gives visitors a comprehensive picture of Tsutomu Kurokawa's career and designs. The timeline format is especially effective for presenting the portfolio to overseas users.

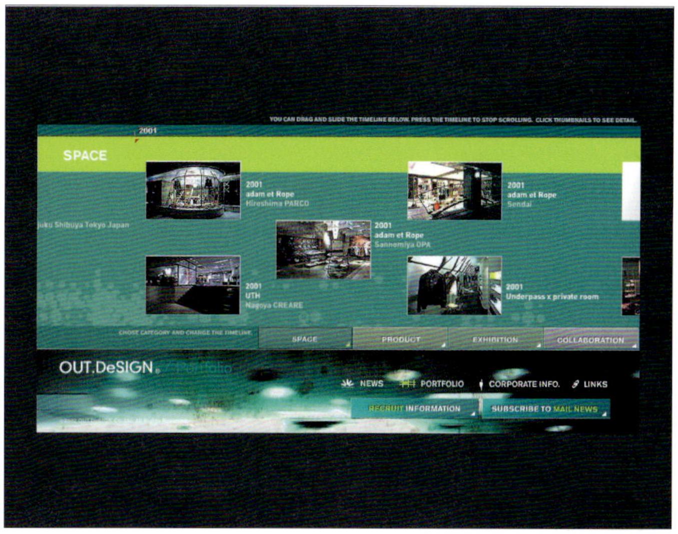

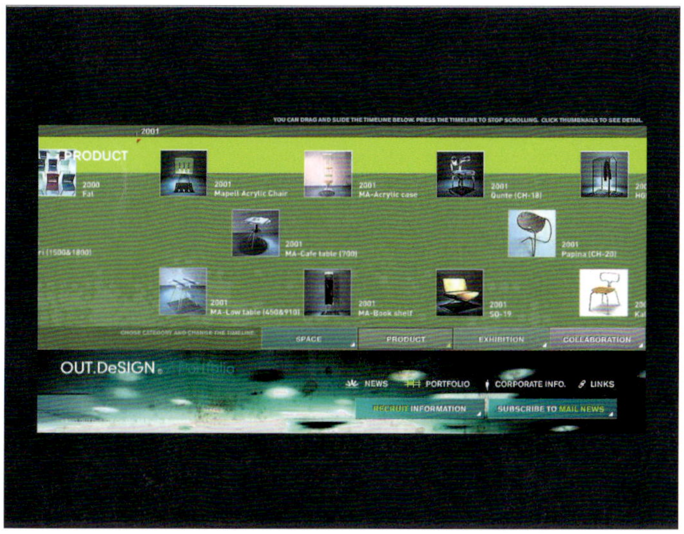

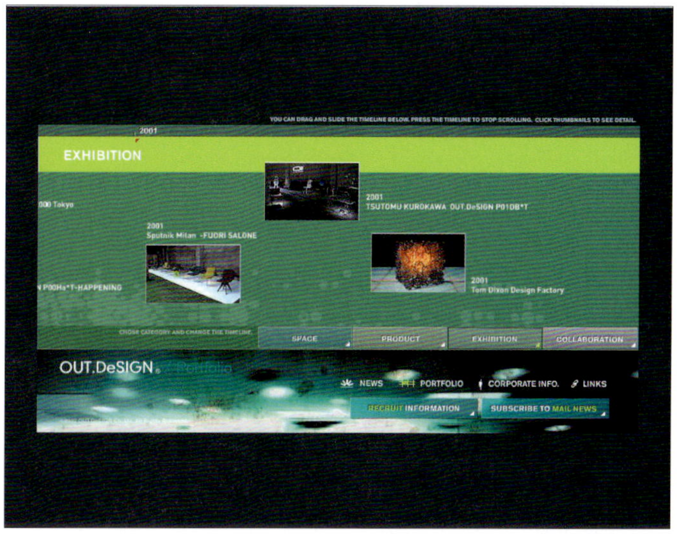

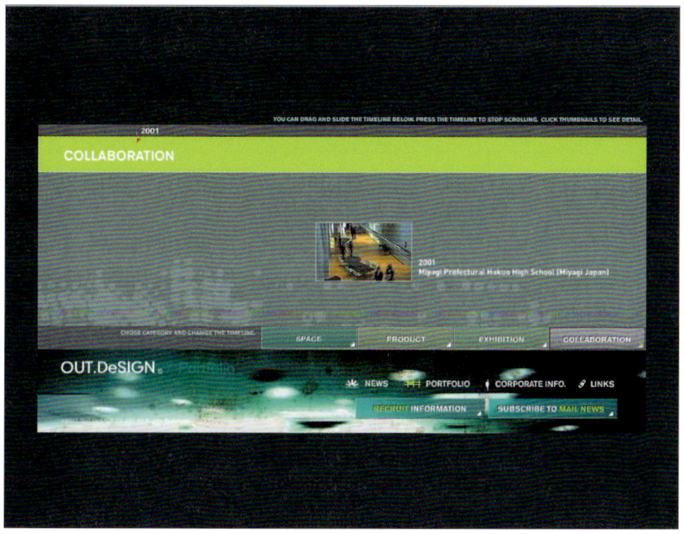

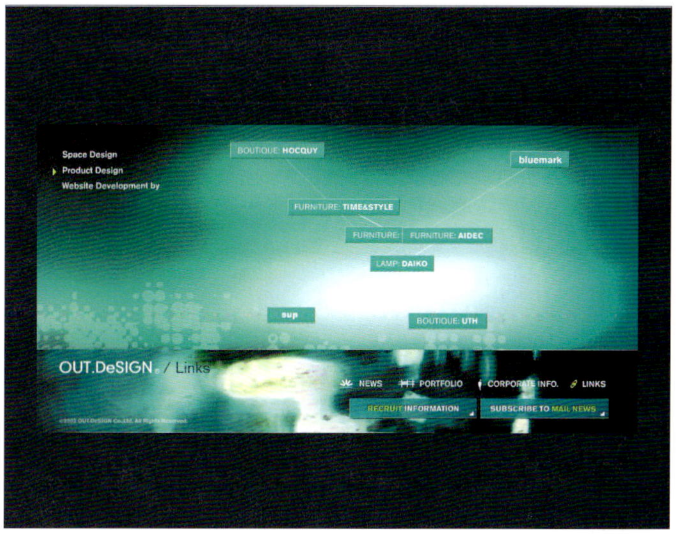

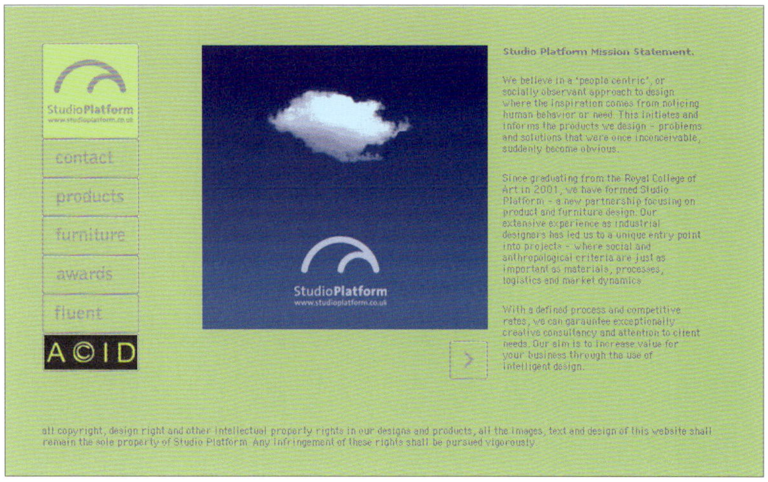

Studio Platform Website **http://www.studioplatform.co.uk**
UK／2001
CL, SB： Studio Platform DF： Nick Jinkinson / Nick Paterson

concept 新進デザイナーコンビ、Nick JinkinsonとNick Patterson（2001年ロンドンの王立芸術学院卒。Ron Aradに師事、製品デザインを学ぶ）のプロモーション用Webサイト。このWebサイトは、クリアかつシンプルをコンセプトとし、Studio Platformが手がける製品デザインの独特なスタイルを伝えるために作成されました。世間で評判のデザイナーが手がける装飾重視の宣伝目的サイトに対抗し、高名より品質ありきであることを実証しています。

The Web site promotes the work of emerging design duo Nick Jinkinson and Nick Paterson. They graduated with Master's degrees from London's Royal College of Art in 2001, after completing the Design Products Course directed by Ron Arad. The site is designed to be clear and simple. Priority was given to communicating Studio Platform's distinctive product design style and manifesto. It is also a reaction to the over-styled and over-hyped sites of some more established design names, and demonstrates that quality must come before super-stardom.

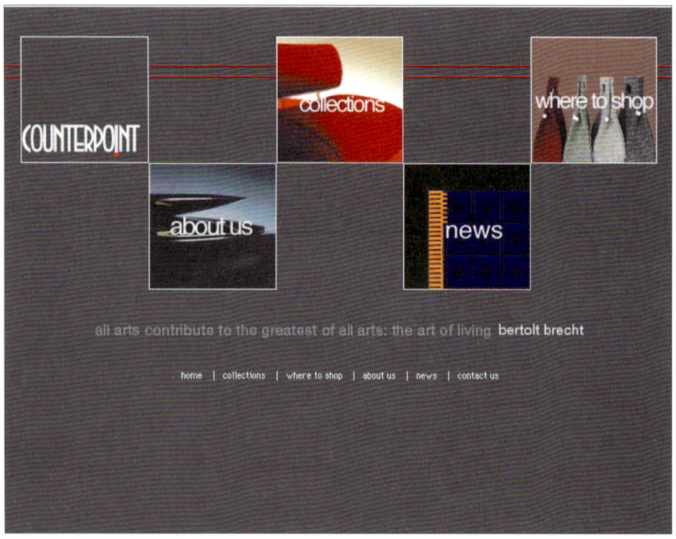

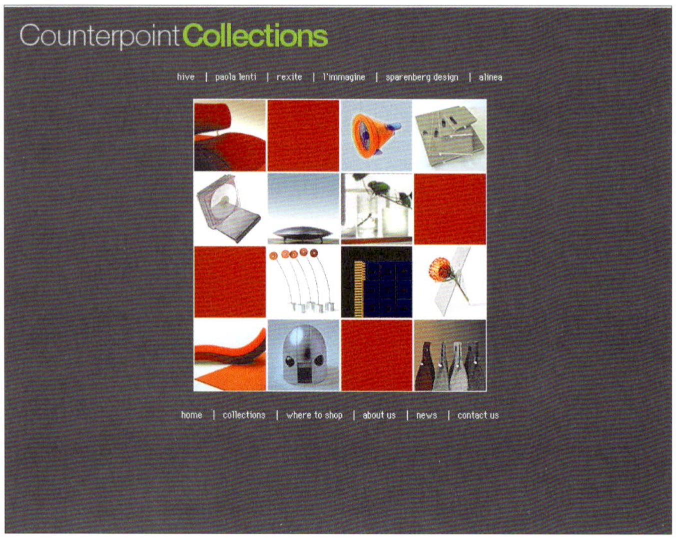

Counterpoint
http://www.counterpoint1.com/
USA／2002
CL, SB： Counterpoint

● **ATYS** ● **http://atysdesign.com** ● **USA／2001**
CL, SB：ATYS Contemporary Living Accessories　DF：Ed Andrews / somelab

concept 世界中の建築デザイン製品の専門店です。取り扱っているのは北欧、ドイツ、イタリア、フランスの近代デザイン製品ですが、国を問わず、新しいレベルの設計に挑戦する優秀な産業デザイナーや建築家に注目しています。Webサイトで紹介されている暖炉用品、計算機、ワイングラス、カフスボタンなどの美しいデザインの家庭用品や個人向けの用品を、オンラインで購入することができます。アートギャラリーや美術館のように製品を見ながら説明を読む感覚は、さながらデザインのタイムカプセルを見ているかのようです。ATYSのWebサイトでは、実際に店舗に足を運ぶのと同じように買い物を楽しむことができます。取り扱い製品が幅広く、分類が難しかったことから、デザイナーのEd Andrewは、新しい製品のデザインを目立たせるために、近代的な独自のフォームと陳列方法を考案しました。グレーのナビゲーションにカーソルを重ねると、製品の一部だけがわずかに表示される、遊び心いっぱいのインタラクティブなスタイルです。このナビゲーションバーは、ATYSの店舗をヒントに、細長いテーブルに製品をランダムにディスプレイするイメージを採用しています。好奇心を駆り立てられた訪問者が製品を手にとる（クリックする）と、その製品の拡大イメージと説明が表示されます。デザインが優れているだけでなく、訪問者の心をくすぐる構成です。

ATYS is a specialty shop featuring architecturally designed products from around the world. We specialize in modern design objects from Scandinavia, Germany, Italy, and France. ATYS focuses on internationally prominent industrial designers and architects who take product design to a new level. Beautifully designed household and personal items as diverse as fireplace tools, calculators, wine glasses, and cufflinks are available for purchase at the site. The art gallery/museum look and detailed descriptions of each product make a visit to the site like a walk through a design time capsule.
ATYS's Web site was designed to give first-time viewers an experience that is similar to actually visiting the store. This was a challenge, because the wide array of products—although they share design excellence—do not fit into neat categories. Designer Ed Andrews of somelab has created an original and modern form and method for product display that stimulates the viewer's interest in the new product designs. Partial views of products that mysteriously slide out of gray navigation bars invite playful interaction. The navigation bar format mimics ATYS's in-store style of displaying a random selection of items on long, narrow tables. More visual information is displayed as visitors demonstrate interest in particular products; when a visitor "picks up" (clicks on) an item to look at it more closely, a detailed product description appears. The site design is architectural and playful.

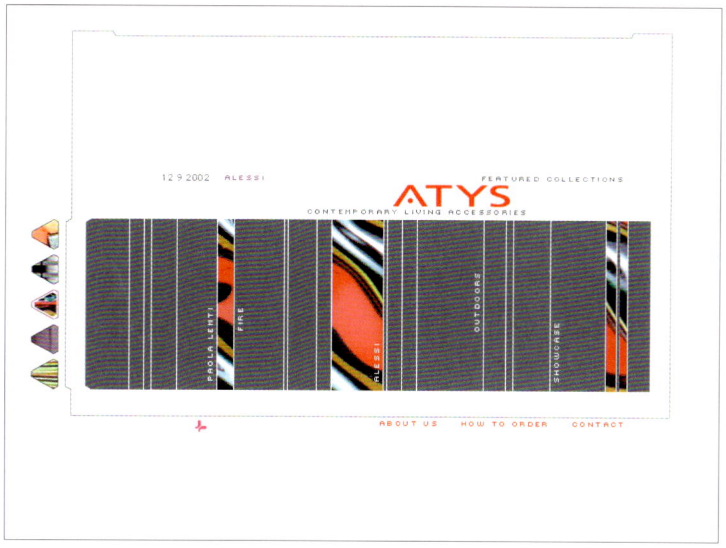

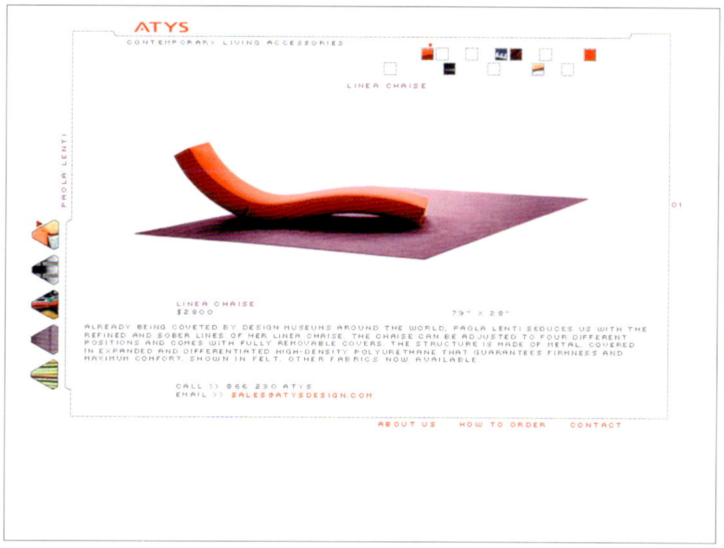

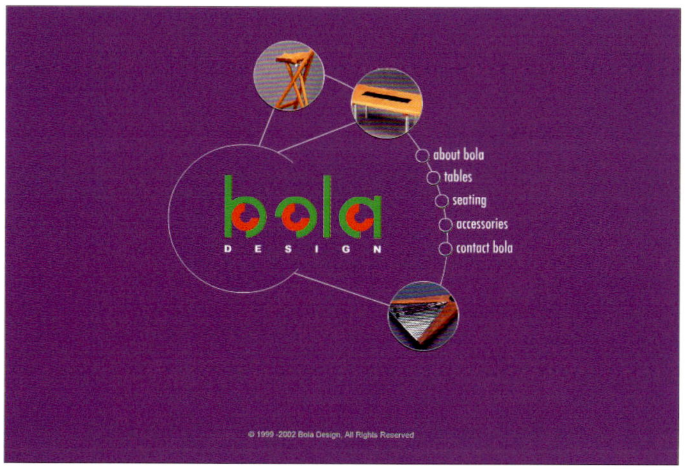

● **bola DESIGN** ● **http://www.bolafurniture.com** ● **USA／2002**
CL, SB：Bola Design

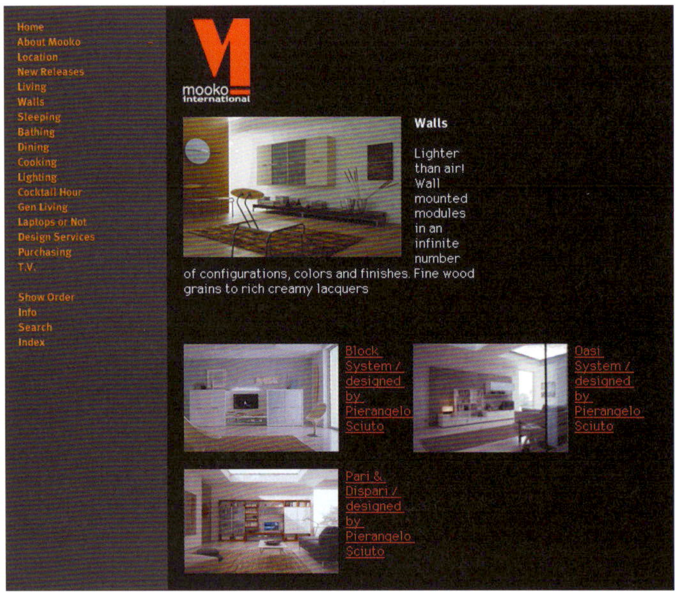

● **Mooko International** ● **http://www.mookointernational.com**
● **USA／1986**
CL, SB：Mooko International Inc. DF：Les Spits

concept ありきたりの家具に満足しないあなたへ
Mookoの家具は、清潔感ある優雅な近代ヨーロッパ式デザインと材料を禅の教えである調和と融合させ、機能美を追求しています。Mookoはお客様自身の表現をお手伝いします。個性を表現するためのパレットでありたいMookoは願っています。機能性と美の双方を徹底して追求した家具の価値が衰えることはありません。どんな空間にも必ず調和します。

Contemporary Furniture for the Uncommon Individual…
Born of a passion for simplicity, Mooko blends the clean, elegant forms of modern European furniture design and materials with the organic symmetry of Zen philosophy. Mooko is a vehicle for your expression. Our purpose is to establish the pallet from which to create one's individual presence. Our timeless designs of both functional and aesthetic form offer our customers lasting value, assuring a space in harmony with itself.

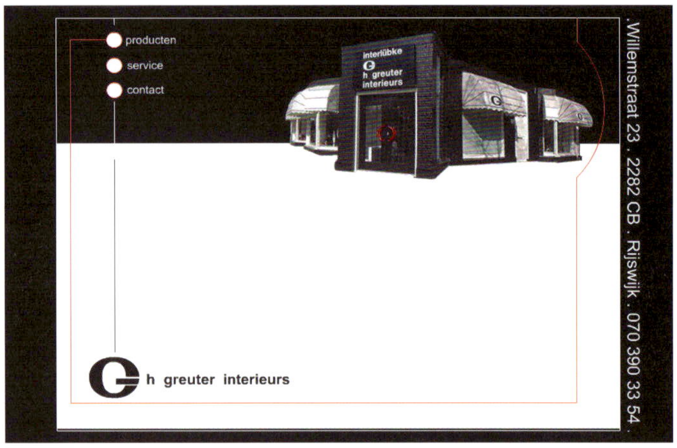

● **Gh greuter interieurs** ● **http://www.interlubke.com/**
● **THE NETHERLANDS／2002**
CL, SB：Gh greuter interieurs

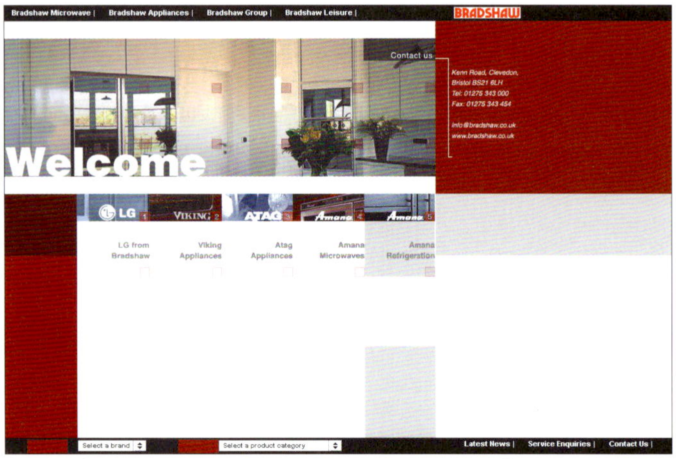

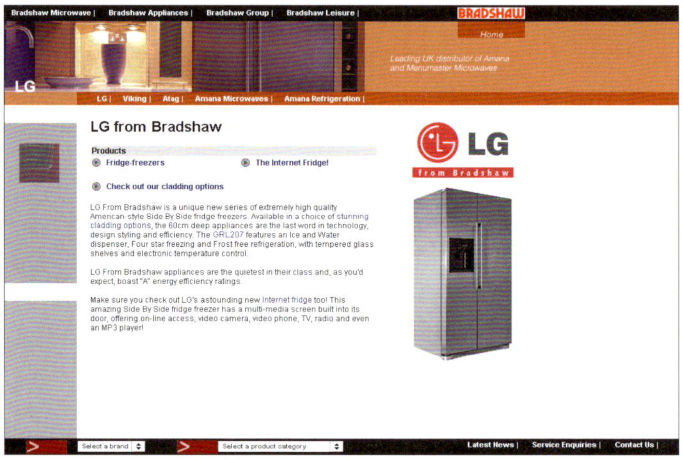

● **BRADSHAW** ● **http://www.bradshaw.co.uk/** ● **UK／2002**
CL, SB：Bradshaw Group

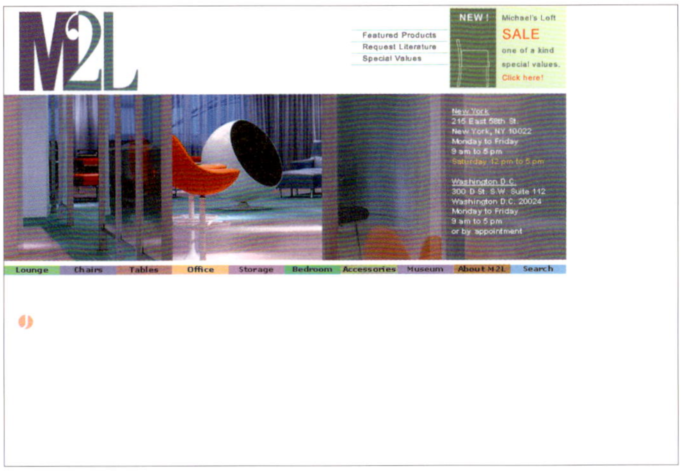

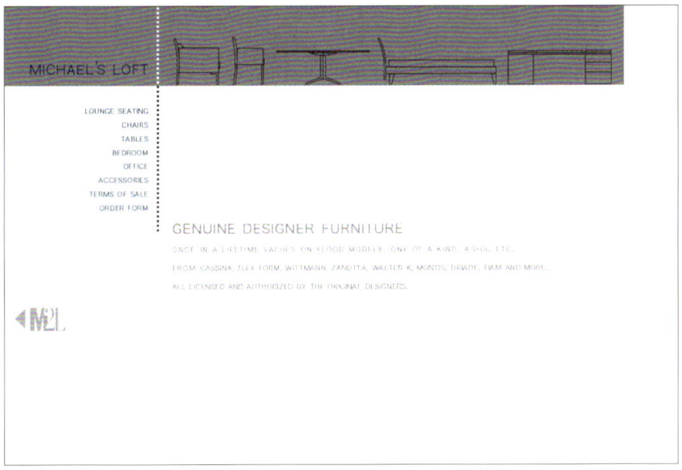

● M2L ● http://www.m2lcollection.com/ ● USA／2002
CL, SB：M2L, Inc.

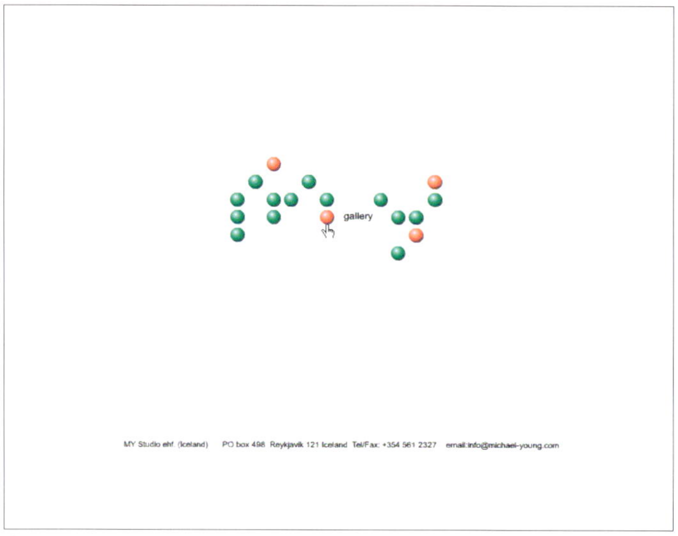

● MY STUSIO EHF ● http://www.michael-young.com ● ICELAND／2002
CL. SB：M.Y.STUDIO EHF DF：WWW.CAOZ.IS

● HELTEN ● http://www.helten.com/ ● USA／2002
CL. SB：HELTEN, BEST OF DESIGN.

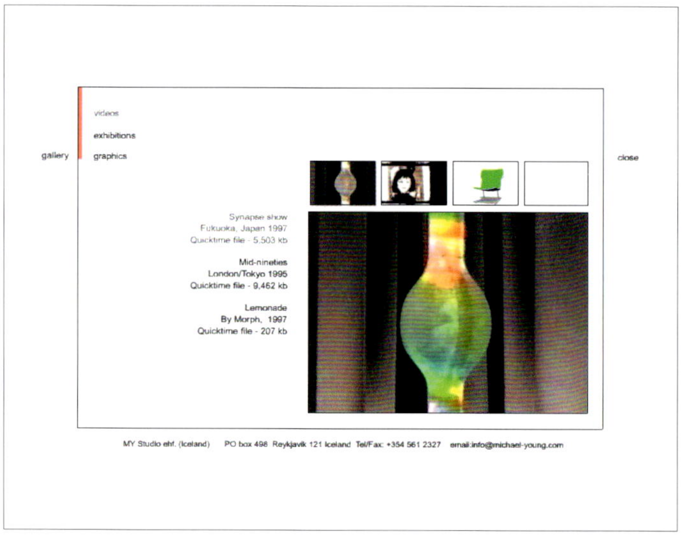

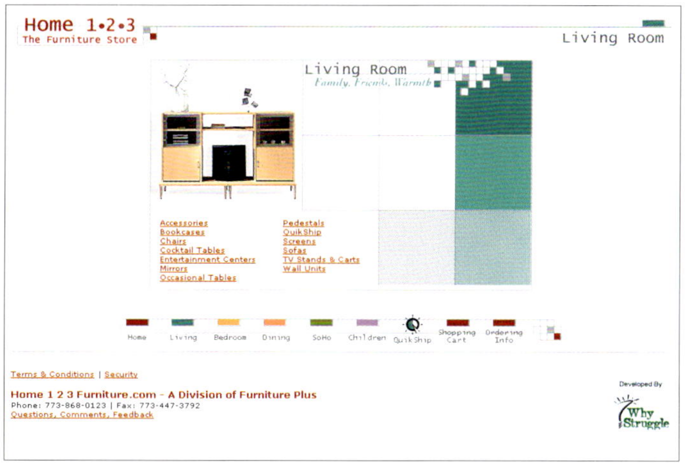

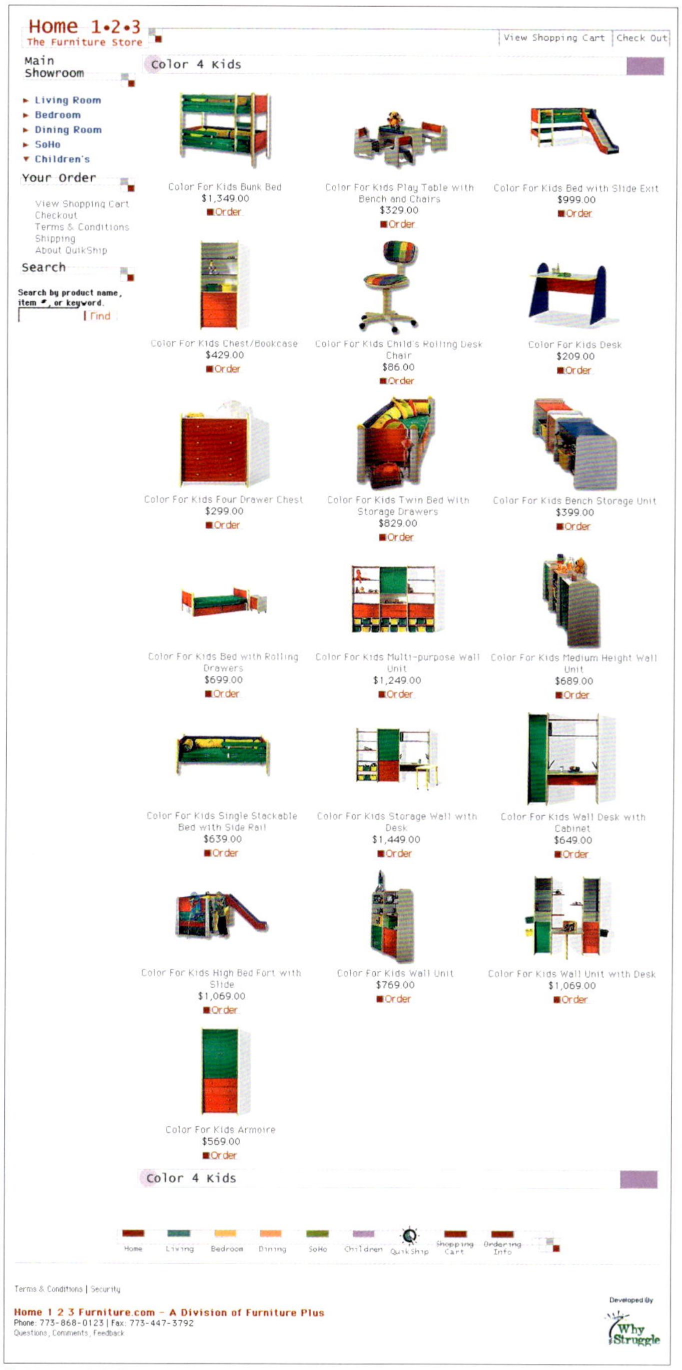

● **Home 1 2 3 - The Furniture Store**
● **http://www.home123furniture.com/**
● **2002**
CL, SB : Home 1 2 3 - The Furniture Store

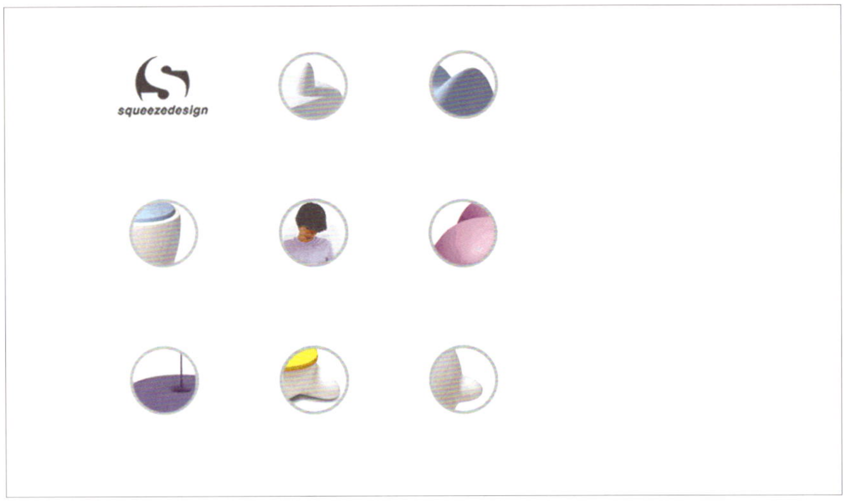

- squeezedesign
- http://www.squeezedesign.com
- SINGAPORE／1999

CL, SB：Squeeze Design　DF：Tristan Tay

concept

すっきりとしたデザインの中に、1つ1つの作品が画面に大きく表示されます。ボタンをクリックして作品をそっとのぞいてみてください。ボタンにはポップな色使いと丸いフォルムが施されています。アニメーション効果を使用したシンプルなトップページを見るだけで、デザイナーの作品の本質である楽しさと幸福感が伝わります。

A special, uncluttered interface was designed to showcase the designer's works by letting each design piece have the screen all to itself. Straightforward navigation uses buttons that give users a sneak peek at the objects. The colors and round shape of the buttons reflect the design objects themselves. Simple animation on the top page conveys the fun, happy nature of the designer's work.

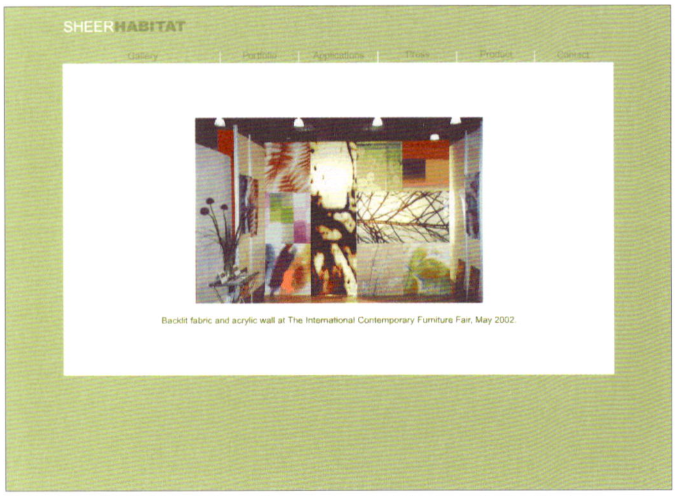

- SHEERHABITAT　● http://www.sheerhabitat.com/　● USA／2002

CL, SB：Sheerhabitat

- Larosa Design & Consulting　● http://www.italianfurnishings.com/
- USA／2002

CL, SB：Larosa Design & Consulting

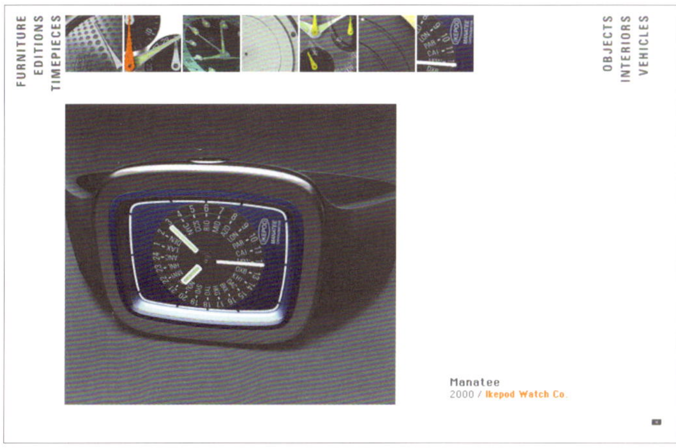

● Marc Newson ● http://www.marc-newson.com/ ● UK／2002
CL, SB：Marc Newson Ltd.

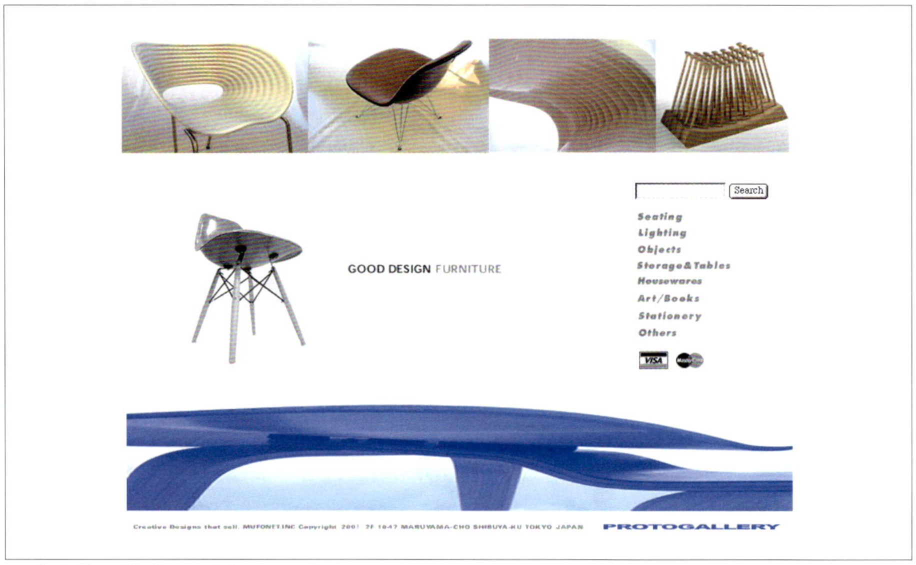

● プロトギャラリー(PROTOGALLERY) ● http://www.protogallery.com/ ● JAPAN／2002
CL, SB：(株)ミューフォネット MUFONET. INC DF：(株)ミューフォネット デザインルーム MUFONET. INC DESIGN ROOM

● TAKIYA ● http://www.takiya.org ● JAPAN／2001
CL, SB：タキヤ(株) TAKIYA CO., LTD. DF：ガウチョ GAUCHO

● NACOM ● http://www.nacom.co.jp ● JAPAN／2001
CL, SB：タキヤ(株) TAKIYA CO., LTD. DF：ガウチョ GAUCHO

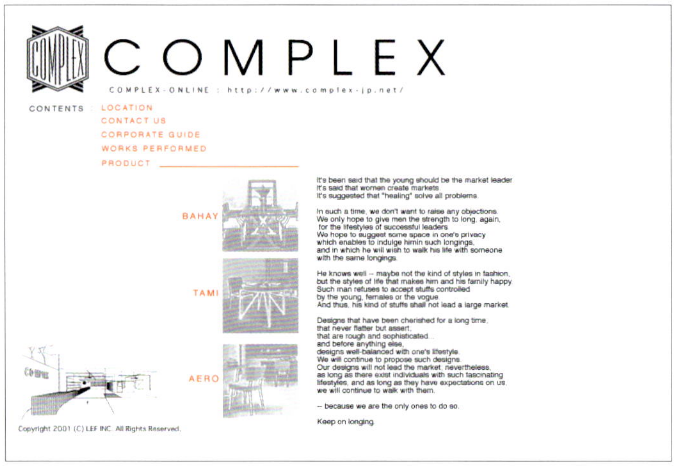

● COMPLEX ● http://www.complex-jp.net/ ● JAPAN／2002
CL, SB：(株)レフ LEF INC. DF：早船裕介 Yusuke Hayafune

● TIMELESS COMFORT ● http://www.timelesscomfort.com
● JAPAN／2001
CL, SB：(株)アスプルンド ASPLUND CO., LTD. DF：サンワ(株) SANWA Co., Ltd.

● **B-COMPANY BUNGO** ● **http://www.b-company.net**
● **JAPAN／2002**
CL, SB：（株）ぶんご　BUNGO CO., LTD.

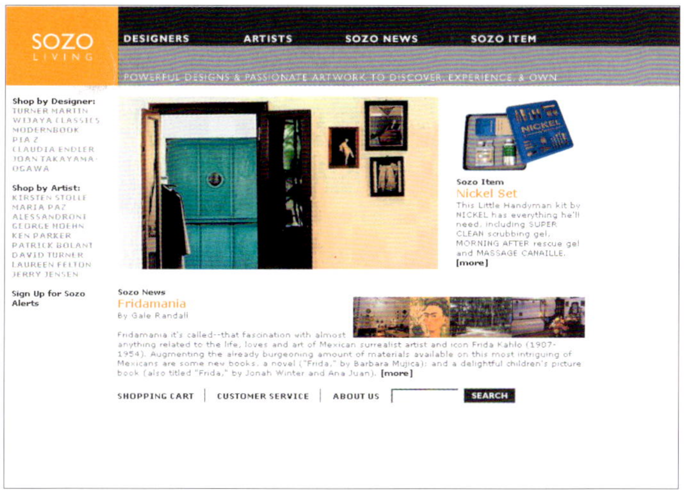

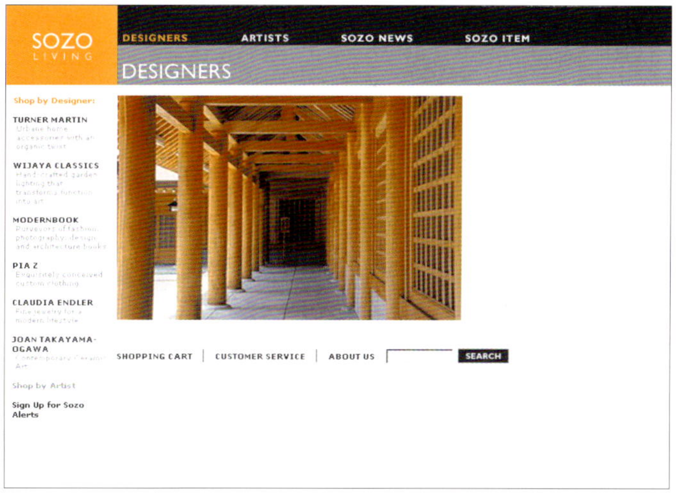

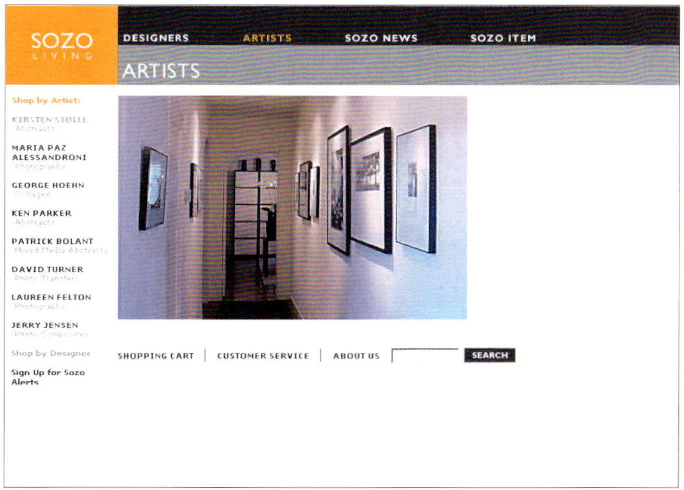

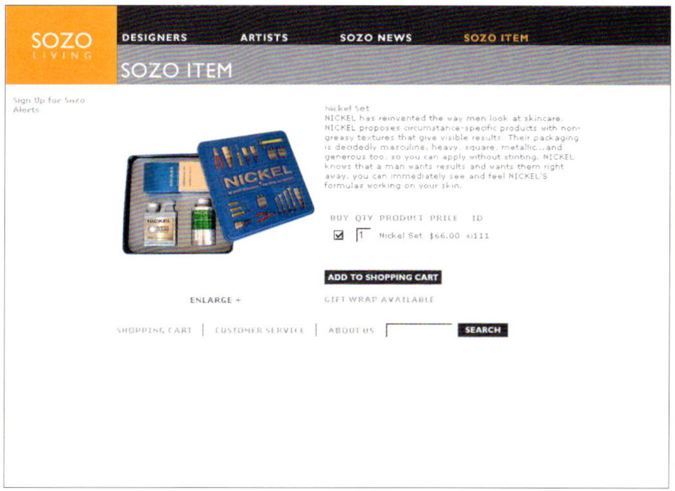

● **SOZO LIVING** ● **http://www.turnermartin.com/** ● **USA／2002**
CL, SB：SozoLiving

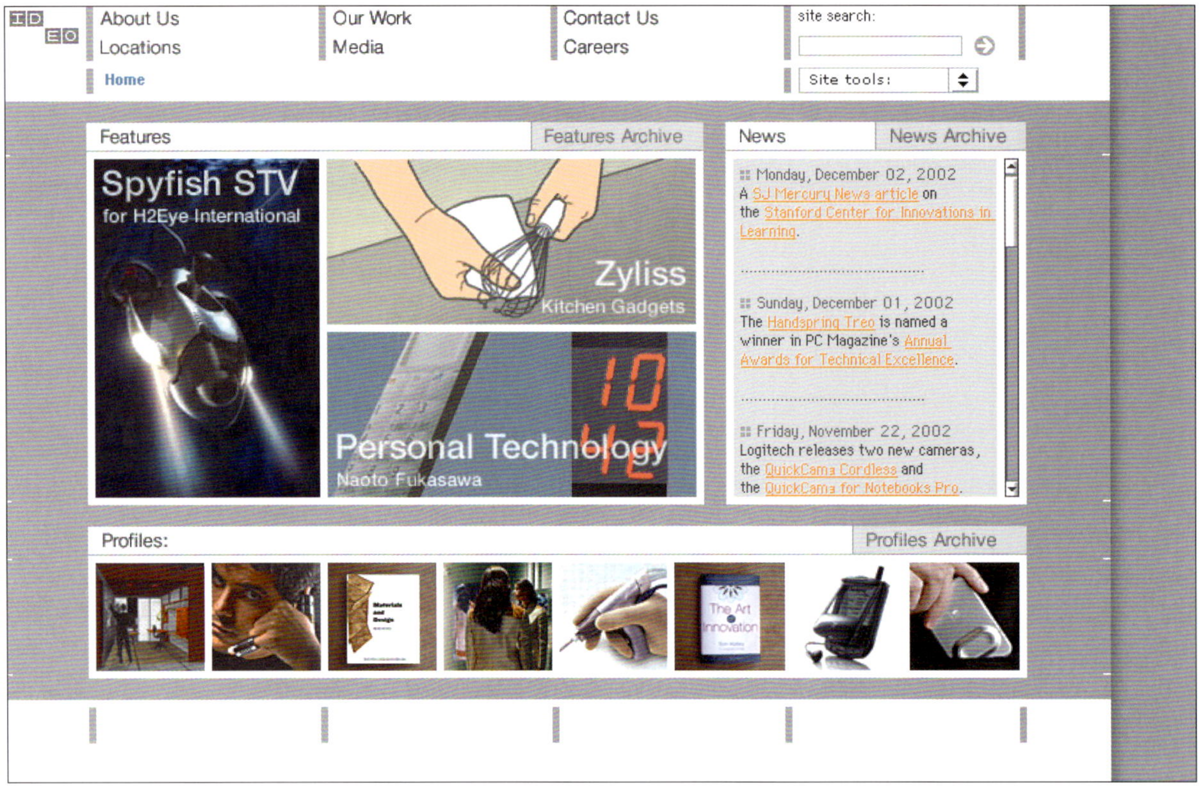

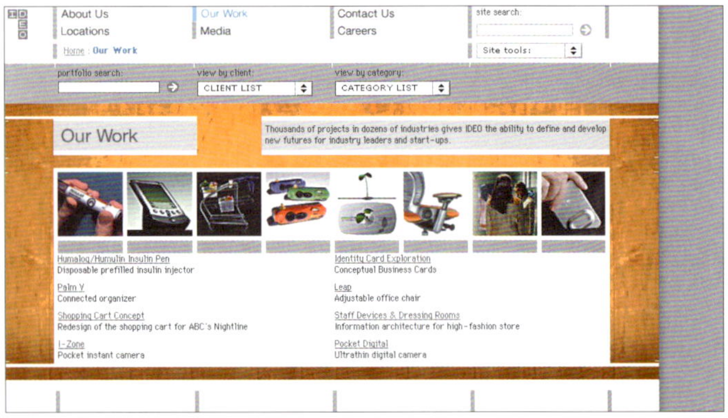

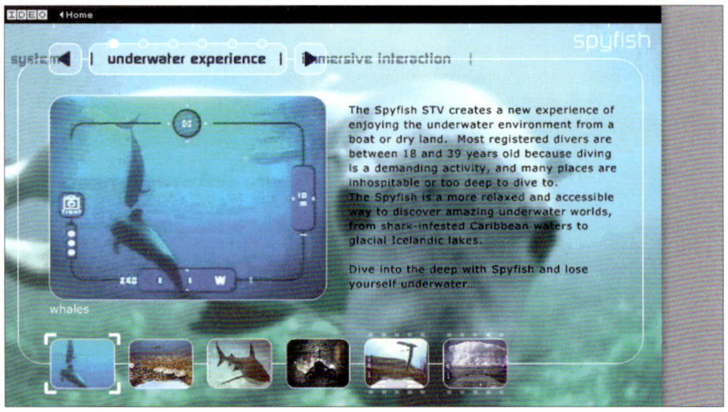

● **IDEO** ● **http://www.ideo.com** ● **USA／2002**

CL, SB：IDEO DF：Danny Stillion / Joseph Kral / Mark Chance / Matt Desmond

concept

以前のWebサイトでは作品の写真を総合的に展示していましたが、特色であるダイナミックカルチャーとコラボレイティブな方法論が反映されていませんでした。2002年6月にリニューアルオープンしたWebサイトの構築では、社内にデザインチームを発足させ、IDEOの人、文化、メソッドを中心に構成しました。

IDEO's former Web site presented a comprehensive picture of the work we do, but did not reflect the dynamic culture and collaborative methodology we are renowned for. The new Web site, relaunched in June 2002, was created by an internal design team to emphasize IDEO's people, culture, and methods.

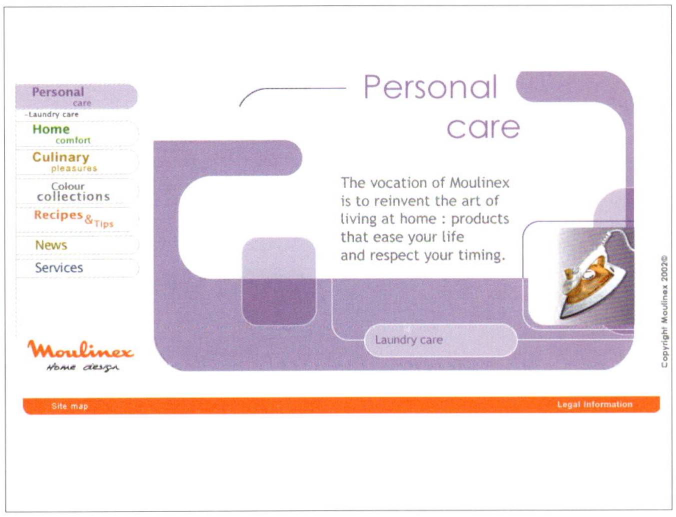

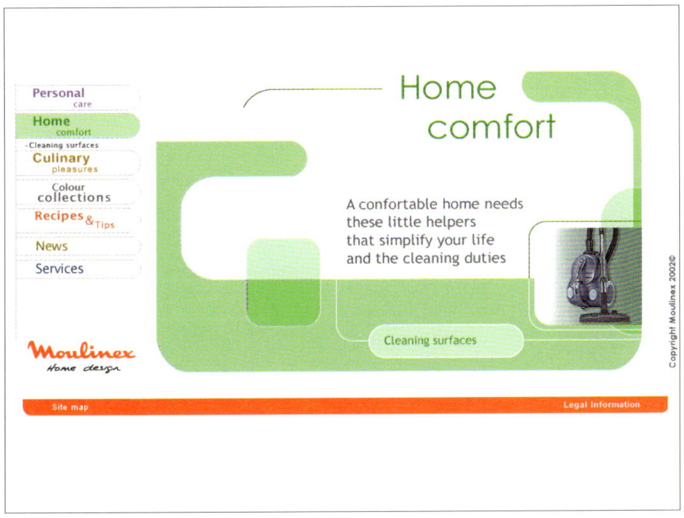

● Moulinex ● http://www.moulinex.com/ ● FRANCE／2002
CL, SB：Moulinex

● BRAUN ● http://www.braun.com
● GERMANY／2001
CL：Braun GmbH SB：Braun DF：www.gft.com (design & site-structure) / www.namics.com (implementation with content management system)

concept Braunの全製品の説明と仕様、および最新情報を掲載。ユーザーフレンドリーな構成とわかりやすいナビゲーションには、Braunが製品開発で追求する機能性と人間工学への熱意が反映されています。Braunのデザイン実績とコアバリューが一致したデザインです。また、品質と革新という要素もデザインに盛り込まれています。訪問者はここで製品のモデルを比較したり、仕様を印刷したりすることができます。「Services」や「Contact」のセクションは充実がはかられ、豊富な情報が掲載されています。

The site has detailed descriptions and specifications on all products, and informational content on the world of Braun. Central to the design are a user-friendly structure and comprehensible navigation that mirror Braun's emphasis on functionality and ergonomics in product design. The site design is consistent with Braun's design credentials and established core values. Other key objectives brought to the design were quality and innovation. Informative overviews permit direct model comparison, profiles of individual products can be effortlessly printed out for all key data, and the Services and Contact sections have been expanded.

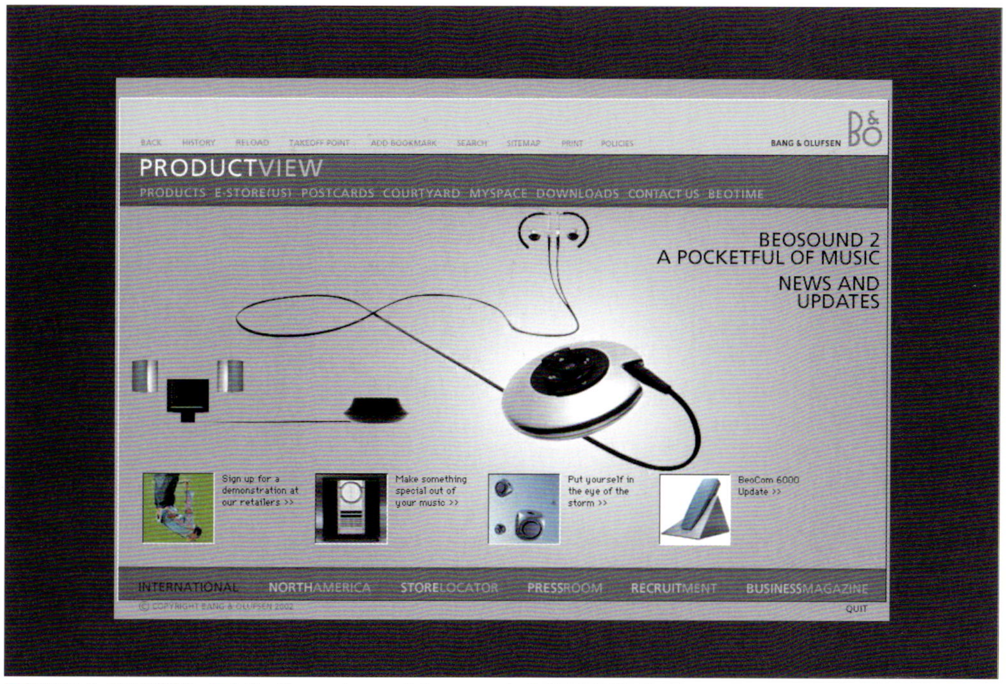

● **BANG & OLUFSEN** ● http://www.bang-olufsen.com/ ● DENMARK／2002
CL, SB：Bang & Olufsen

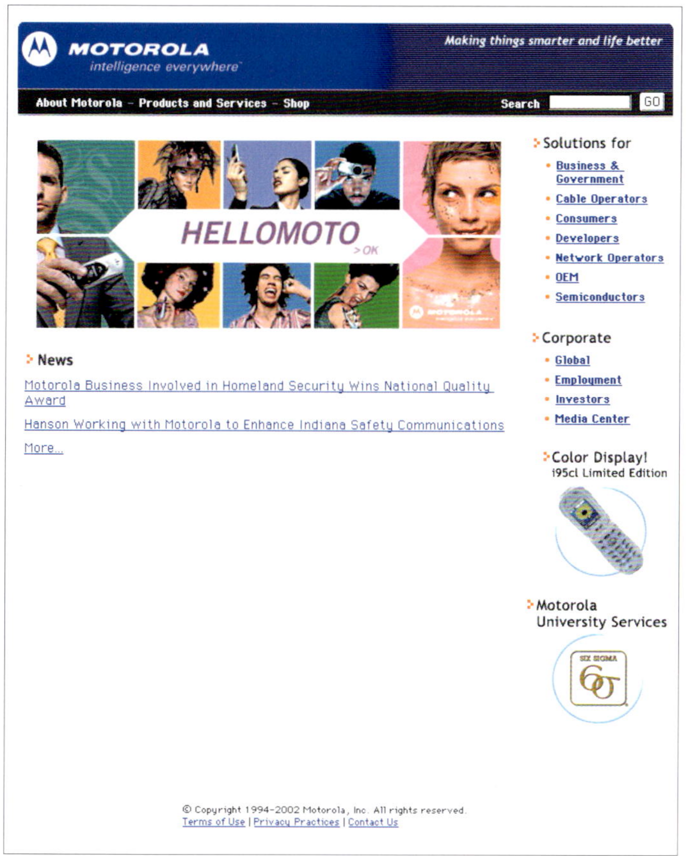

● **MOTOROLA** ● http://www.motorola.com/ ● USA／2002
CL, SB：Motorola, Inc.

● **Trade Air System Ltd.** ● http://www.tradeair.com.hk/ ● CHINA／2002
CL, SB：Trade Air System Ltd.

● **Electrolux** ● http://na.electrolux.com/ ● USA／2002
CL, SB：The Electrolux Group.

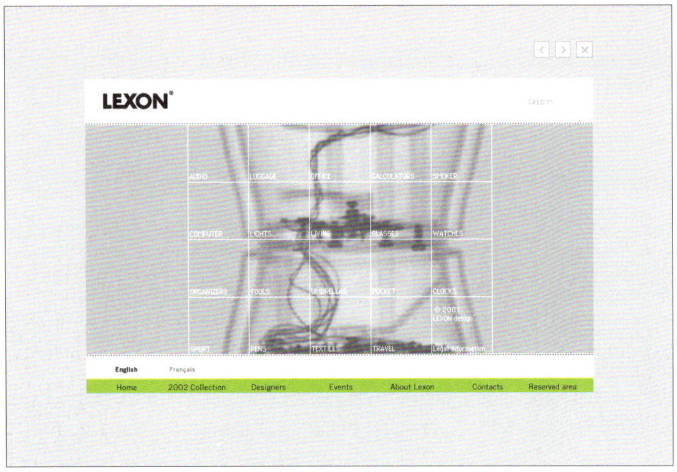

● **LEXON** ● **http://www.lexon-design.com/** ● **FRANCE／2002**
CL, SB：LEXON

● **LA CIE** ● **http://www.lacie.co.jp/** ● **JAPAN／2002**
CL, SB：ラシージャパン（株） LaCie Japan K.K. DF：（株）ペリテックインターナショナル Peritech International Inc.

● **ADTEC** ● **http://www.adtec.co.jp/** ● **JAPAN／2001**
CL, SB：（株）アドテック ADTEC Co., Ltd. DF：（株）インフォバーン Info bahn Inc.
（株）DNPデジタルコム DNP DIGITALCOM Co., Ltd.

● **Adobe** ● **http://www.adobe.co.jp/store/main.html**
● **JAPAN／2002**
CL, SB：アドビ システムズ（株） Adobe Systems Co., Ltd.

● **LE KLINT** ● **http://www.leklint.com/** ● DENMARK／2002
CL, SB : Le Klint

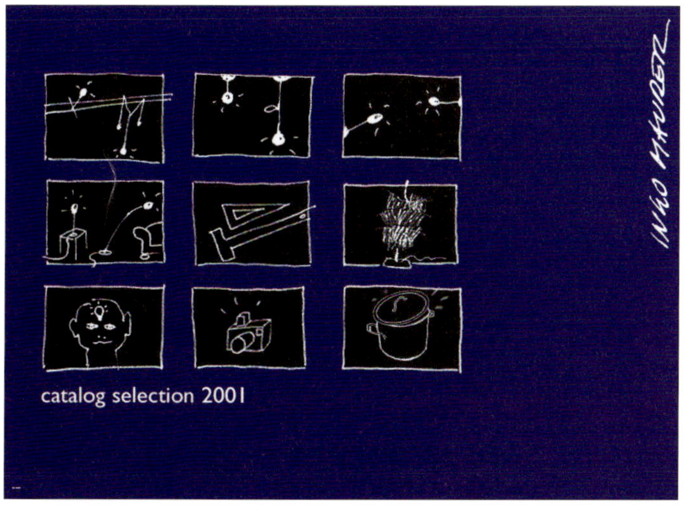

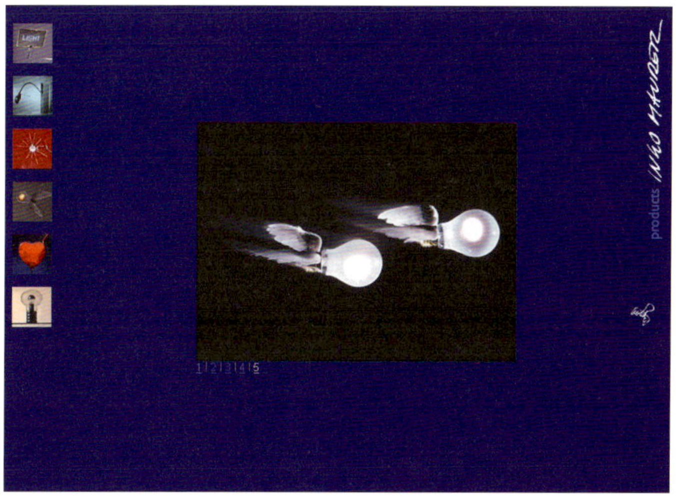

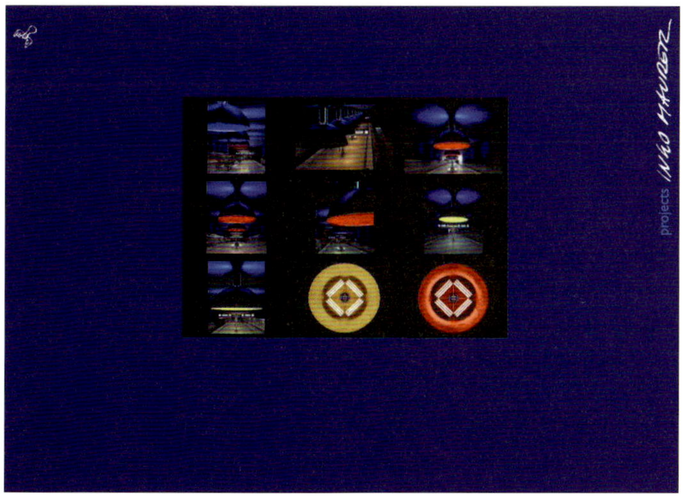

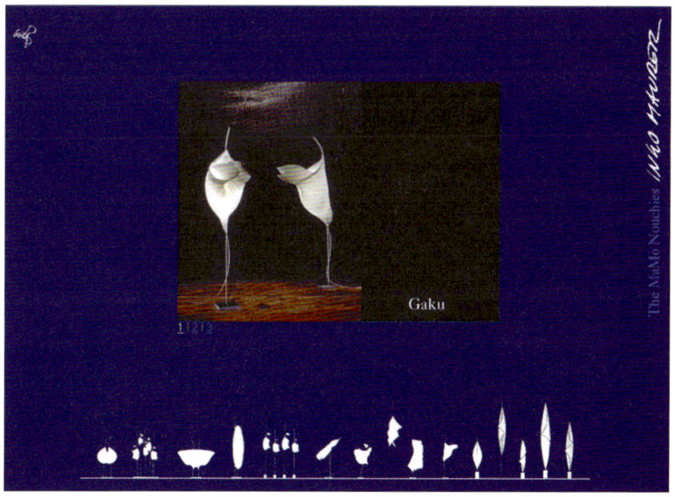

● **Ingo Maurer** ● **http://www.ingo-maurer.com** ● GERMANY／1999
CL : Ingo Maurer SD : Ingo Maurer GmbH DF : Mr. Ingo Maurer / Mr. Hagen Sczech / Mrs. Flavia Thumshirn / Mr. Lino Wiehen, Munich (GER)(1999 "Activ8") / Wiesbaden (GER)(2002 "Scholz & Volkmann")

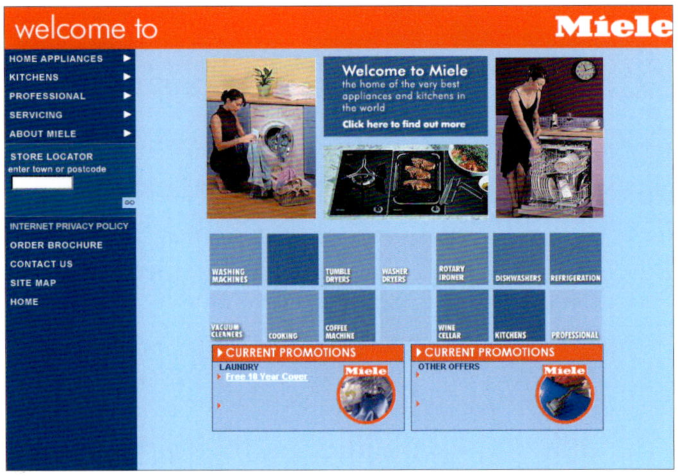

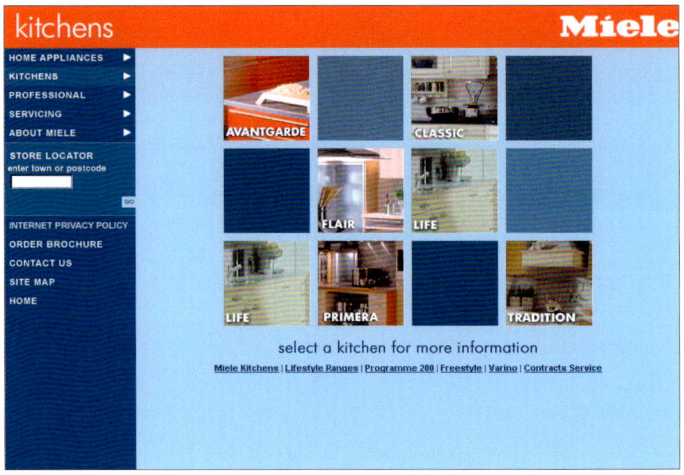

● Miele ● http://www.miele.co.uk/ ● GERMANY／2002
CL, SB： Miele Company Limited

● BOSCH ● http://boschappliances.com ● USA／2002
CL： BSH Home Appliances Corporation SB： Blue C Interactive DF： Marcel Seva

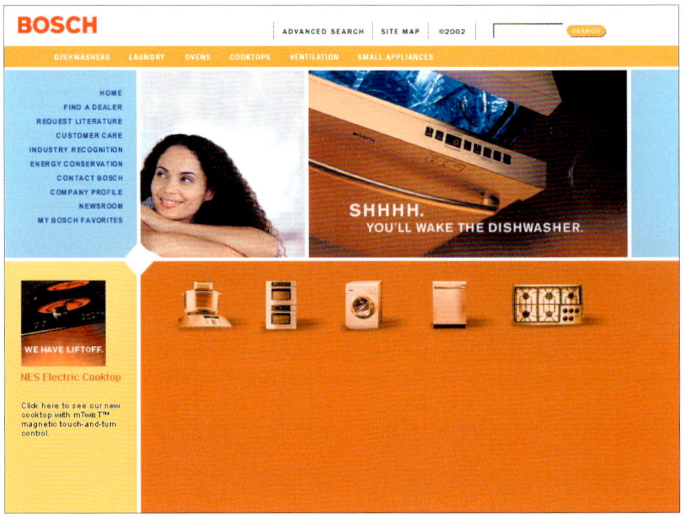

● KitchenAid ● http://www.kitchenaid.com/ ● USA／2002
CL, SB：KitchenAid

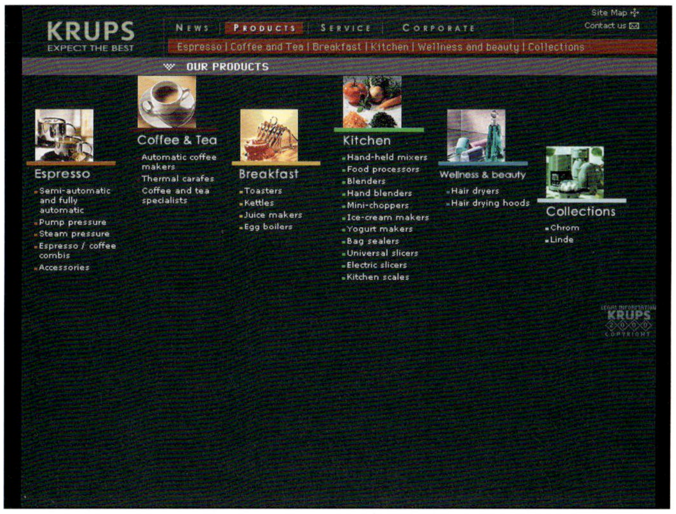

● KRUPS ● http://www.krups.com/ ● GERMANY／2002
CL, SB：Groupe SEB

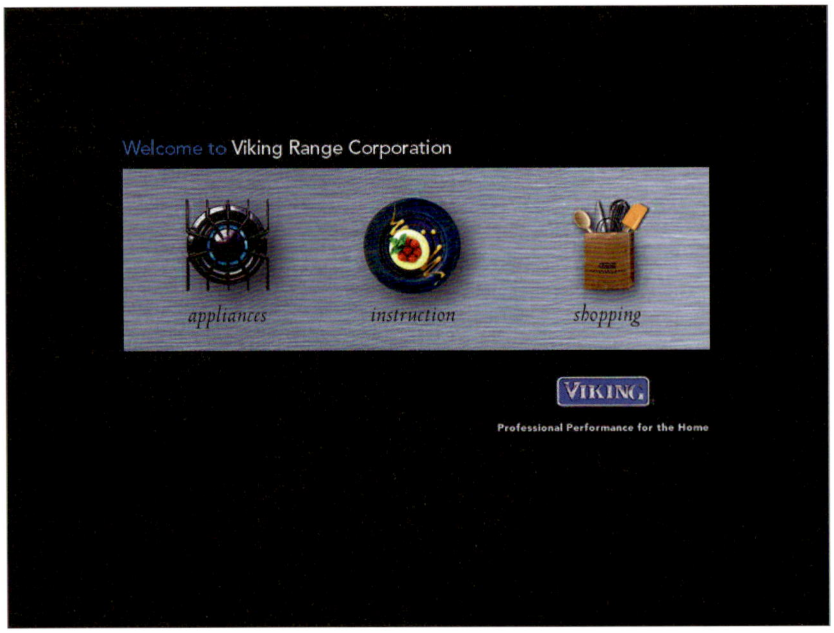

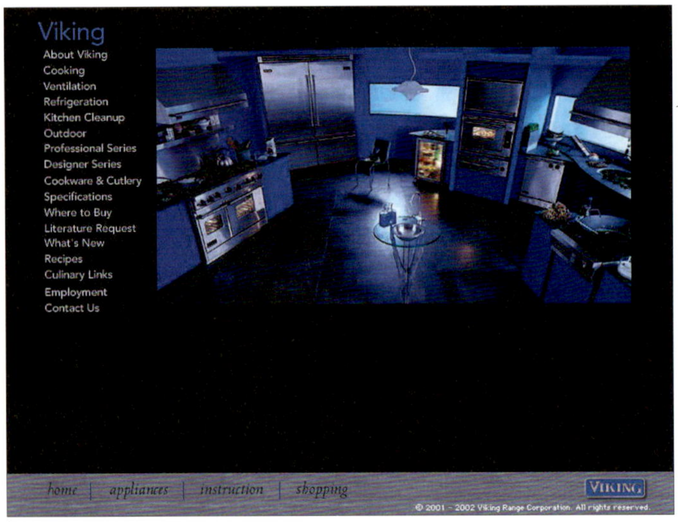

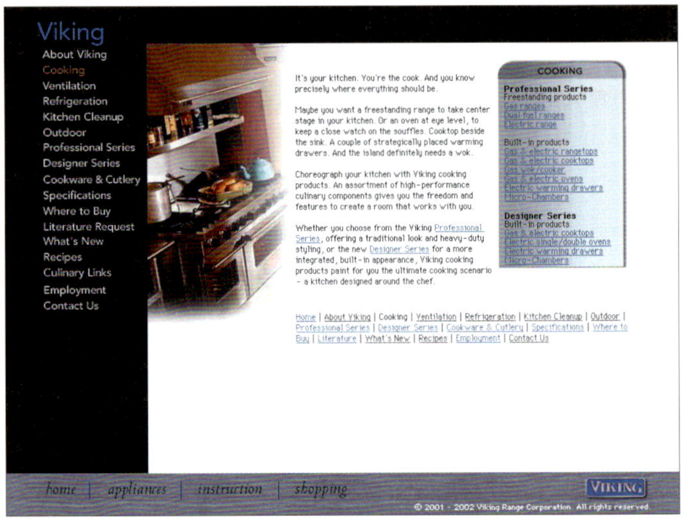

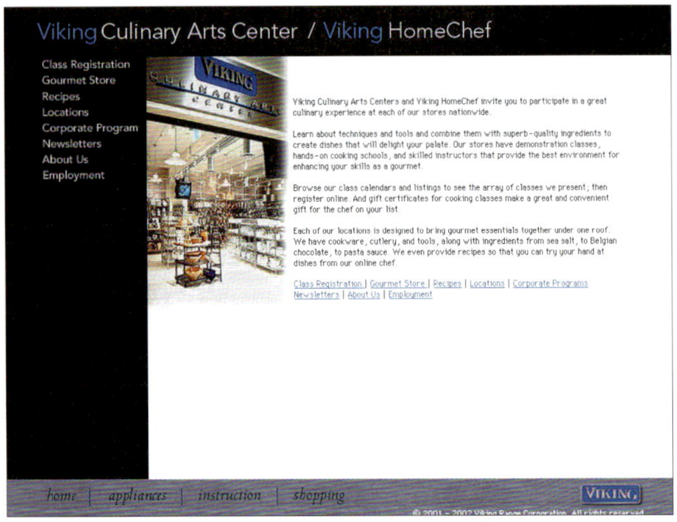

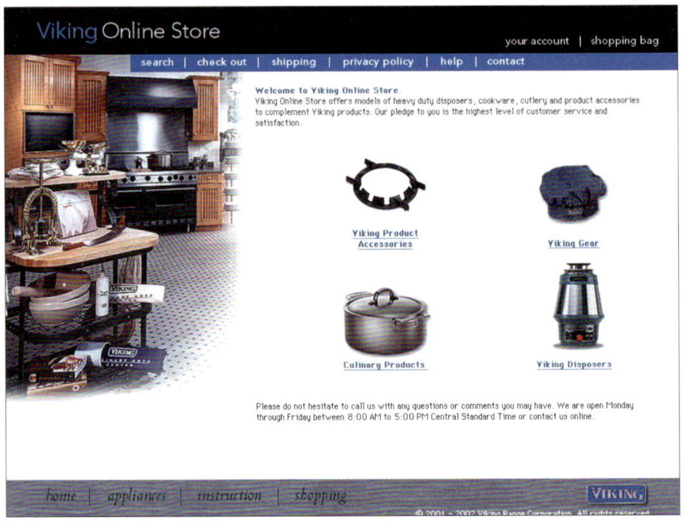

● VIKING ● http://www.vikingrange.com/ ● USA／2002

CL, SB： Viking Range Corporation

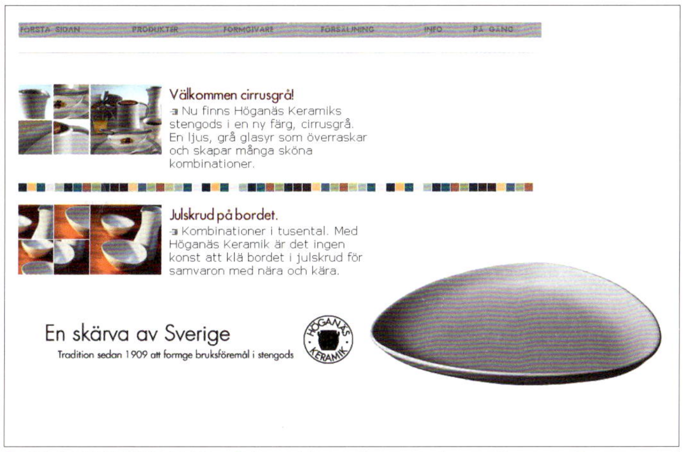

● HOGANAS KERAMIK ● http://www.hoganaskeramik.com ● SWEDEN／2001
CL： Trick Interaction, Helsingborg, Sweden SB： BodaNova-Hoganas Keramik AB DF： Lotta Morberg

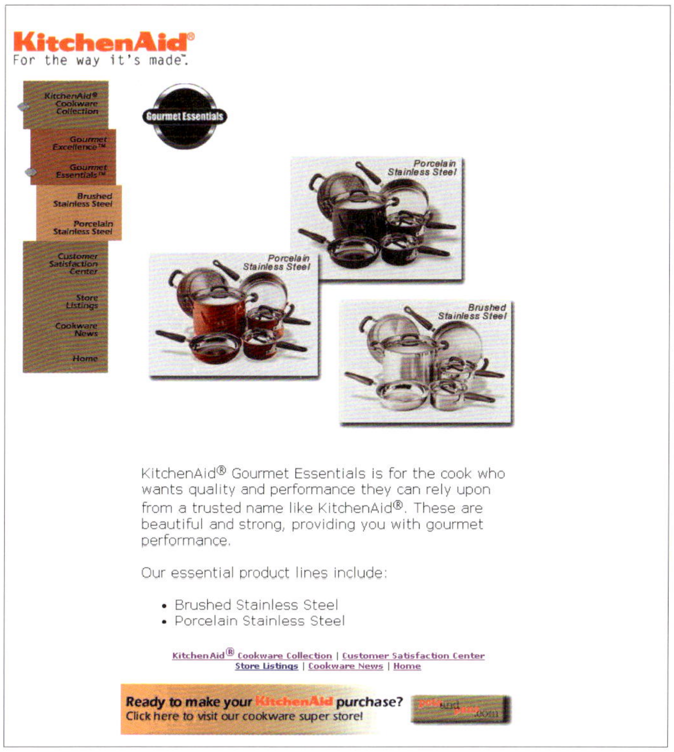

● KitchenAid ● http://www.kitchenaidcookware.com/ ● USA／2002
CL, SB： KitchenAid

● **無印良品（MUJI）** ● **http://www.muji.net/** ● **Japan／2000**
CL：良品計画／ムジ・ネット（株） Ryohin Keikaku Co., Ltd. / MUJI.net Co., Ltd.
SB, DF：ムジ・ネット（株） MUJI.net Co., Ltd.

concept 「もっと深く、もっと広がる無印良品の世界」をテーマに、無印良品のe-コマースや新しい事業を展開しています。ネットコミュニティーでは、お客様と共にWEB上の商品開発も行っていて、「持ち運びできるあかり」など新しい商品が生まれています。

"The Expanding World of Muji" is the theme of our Web site. Our goal is to develop our e-commerce business and other new businesses. We are also working on product development together with customers in the Muji net community; a portable lamp was the first of several popular new products this effort has produced to date.

● 鳩居堂（KYUKYODO） ● http://www.kyukyodo.co.jp/ ● JAPAN／2002
CL, SB：（株）京都鳩居堂 KYOTO KYUKYODO CO., LTD. ／（株）東京鳩居堂 TOKYO KYUKYODO CO., LTD. DF：（株）アニマックス ANIMAX, INC.

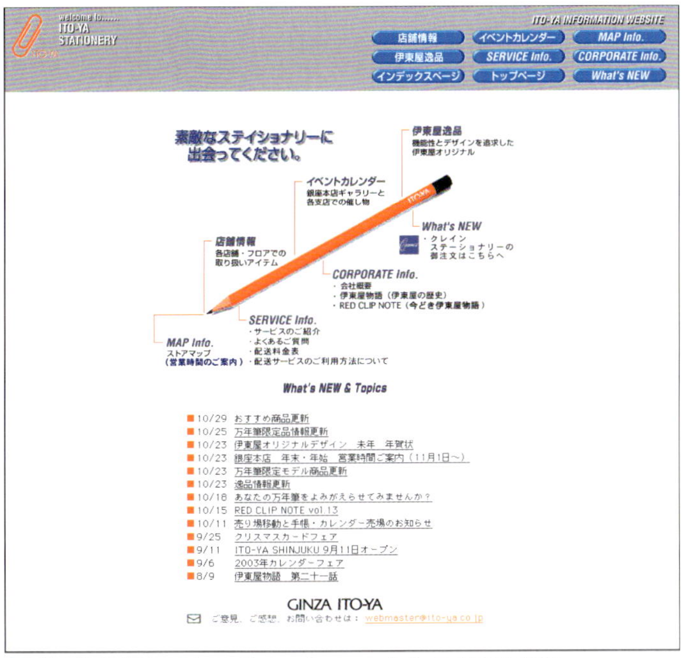

● 伊東屋（ITO-YA） ● http://www.ito-ya.co.jp/ ● JAPAN／2002
CL, SB：（株）伊東屋 Ito-Ya Ltd. DF：（株）ミツエーリンクス Mitsue-Links Co., Ltd.

● ロメオ（Romeo） ● http://www.e-romeo.info/ ● JAPAN／2002
CL：ロメオ Romeo SB：銀座・伊東屋 GINZA ITO-YA DF：（有）セオ 佐藤効省 seo, inc. Kosei Sato／ラフィーノ・デザイン キクヤヨシミ rafino.design Yoshimi Kikuya／アートレス（有）佐藤聡一郎 artless, inc. Soichiro Sagawa

● **ARANZI ARONZO** ● http://www.aranziaronzo.com/ ● JAPAN／1998
CL, SB：(有)アランジアロンゾ ARANZI ARONZO INC. DF：marble 梶原綾子 marble Ayako Kajiwara

concept アランジアロンゾの世界観を忠実に表現するように心がけて制作しています。2002年5月から海外向けオンラインショップがスタートし、世界中にアランジアロンゾの商品をお届けできるようになりました。同年に始まった「てつの部屋」では、ほぼ毎日アランジアロンゾで働く犬てっちゃんをアランジアロンゾさんご本人がレポートしています。アランジネットでやってみたいことを今後も少しづつ実現していく予定です。

Special care was taken in creating this site to remain true to the Aranzi Aronzo world. With the opening of an online shop in May 2002, it has become possible to send Aranzi Aronzo products around world. Also started this year is a daily report on a dog that "works" at Aranzi Aronzo. The dog's name is Tetsu, and the report is labeled Tetsu no heya [Tetsu's Room]. Plans are underway for additional projects on the Aronzo Net site.

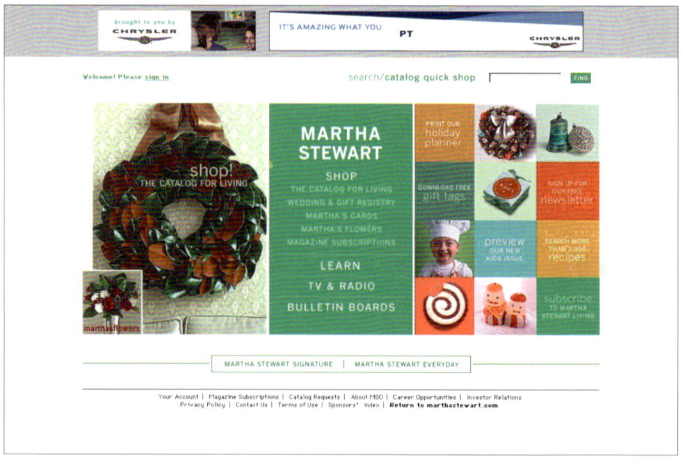

● **MARTHA STEWART** ● http://www.marthastewart.com/ ● USA／2002
CL, SB：Martha Stewart Living Omnimedia

● marimekko ● http://www.txtlart.com/ ● USA／2002
CL, SB：TEXTILE ARTS

● Mr. Friendly ● http://www.mrfriendly.co.jp ● JAPAN／2002
CL, SB, DF：(株)スーパープランニング Super Planning Co., Ltd.

concept 弊社、オリジナルキャラクター「MR.FRIENDLY」のウェブサイトです。キャラクターの「世界のひとびとにハッピーを届けたい！」というコンセプトやグッズの紹介、直営店「MR.FRIENDLY DAILY STORE」についてインフォメーションする場として開設しています。また、このサイトの会員になっていただくことにより、オリジナルスクリーンセーバーやアイコンの無料配信や掲示板などもご利用いただけます。POPなカラー使いや、静止画だけでなく動きのあるページにしています。

This is the web site of Mr. Friendly, the Super Planning character. Introduced are character goods in keeping with Mr. Friendly's objective to bring happiness to all the people in the world. The site also provides information on the Mr. Friendly Daily Store, a 100% Super Planning venture. Members have access to free original screen savers and icons as well as a bulletin board. The site is characterized by its pop-art colors and by its combination of static and moving illustrations.

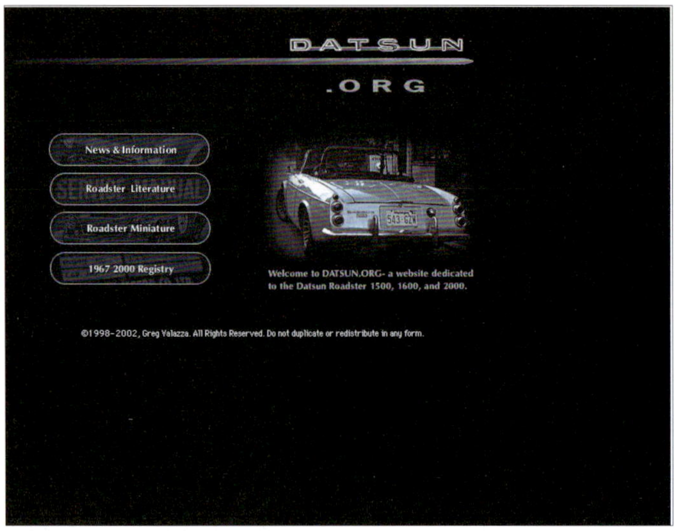

● **DATSUN. ORG** ● **http://www.datsun.org/roadster/default.html**
● **USA／1998-2002**
CL, SB, DF：Greg Valazza

concept DATSUN.ORG は、Datsun Roadster 1500, 1600, 及び2000を特集した非営利websiteです。このサイトは、CARS.COMの1999年度site of the yearを受賞しました。

DATSUN.ORG is a non-profit website dedicated to the Datsun Roadster 1500, 1600, and 2000. It is the winner of the 1999 CARS.COM Site of the Year award.

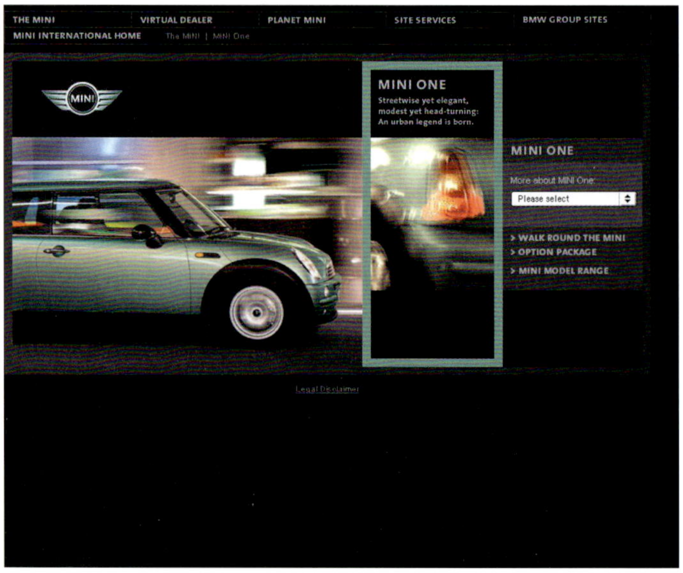

● **MINI** ● **http://www.MINI.COM/** ● **GERMANY／2002**
CL, SB：BMW AG, Munich Germany

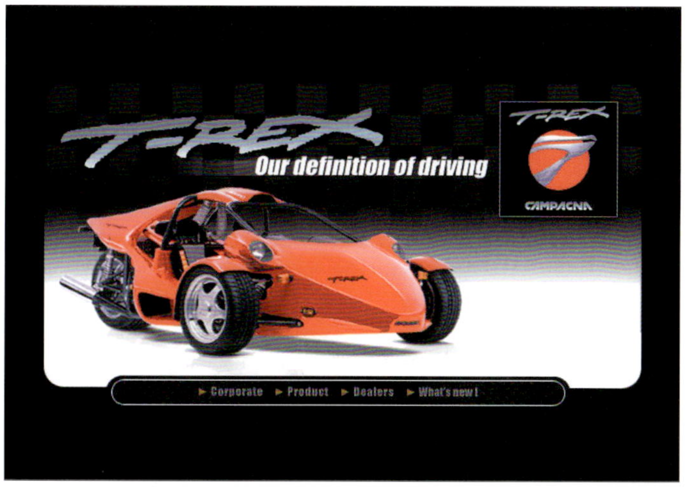

● T-REX ● http://www.go-t-rex.com ● CANADA／2002
CL, SB：Campagna Moto Sport inc. DF：Studio Virtuel Concept

concept サウンド効果、派手なレーシングカラー、チェッカーフラッグをバックグラウンドに使用し、レースの雰囲気を味わえるT-RexのWebサイト。T-Rexのエンジン音が鳴り響き、レーシングカーのスピードとパワーを感じることができます。

The site uses sound effects, bright racing colors, and a checkered flag background design to create a racing atmosphere in which to display the T-Rex. The intro starts off with the T-Rex engine sound to give visitors an idea of the vehicle's speed and power.

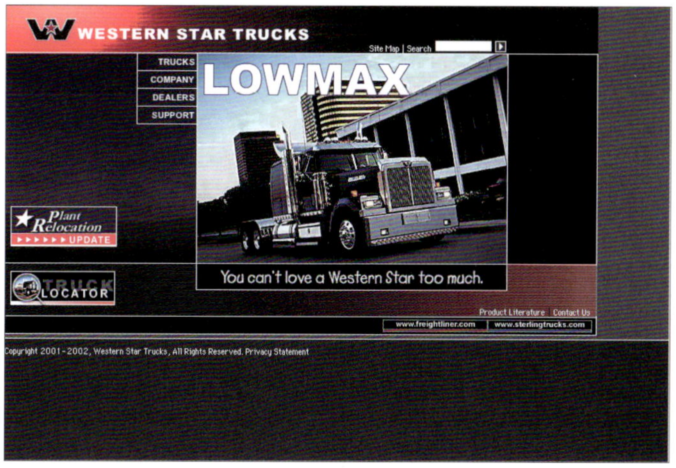

● Western Star Trucks ● http://www.westernstartrucks.com/
● USA／2002
CL, SB：Western Star Trucks

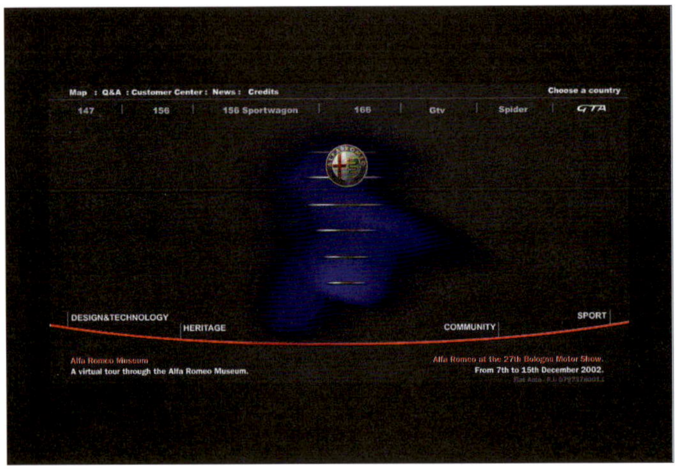

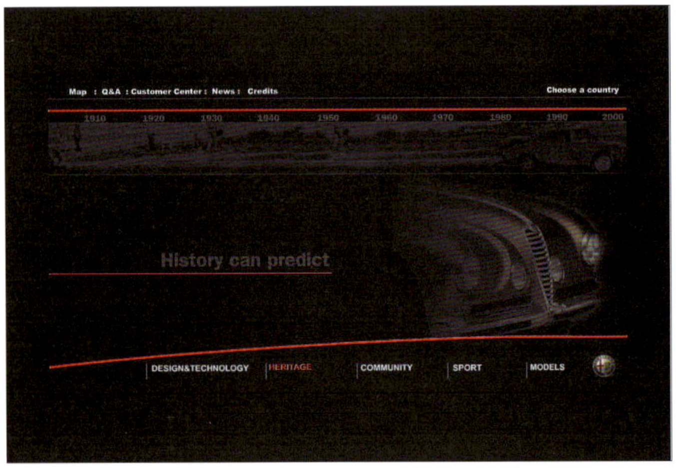

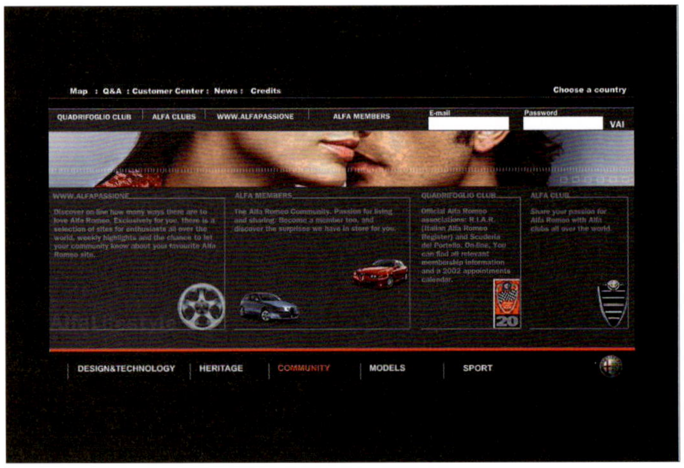

● **Alfa Romeo** ● **http://www.alfaromeo.com/** ● **ITALY／2002**

CL, SB : Fiat Auto S.p.A.

● **Lamborghini** ● **http://www.lamborghini.com/** ● **ITALY／2002**
CL, SB：Lamborghini

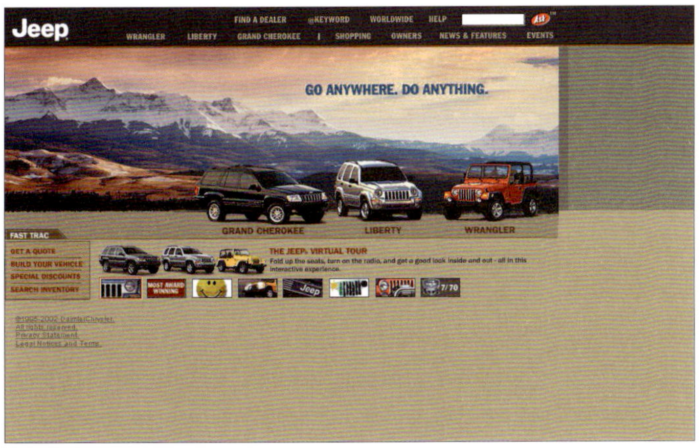

● **Jeep** ● **http://www.jeep.com/** ● **USA／2002**
CL, SB：DaimlerChrysler.

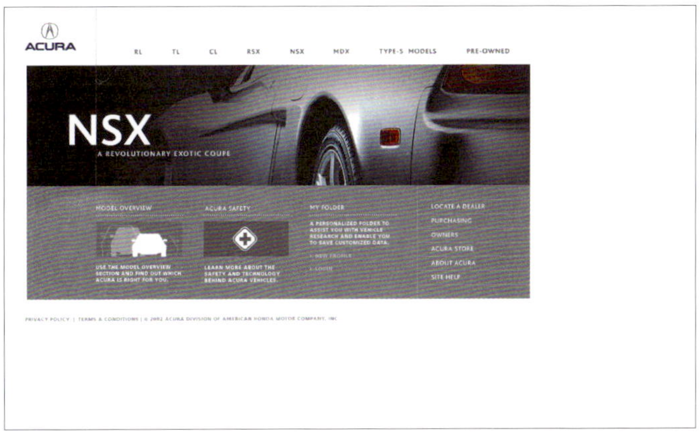

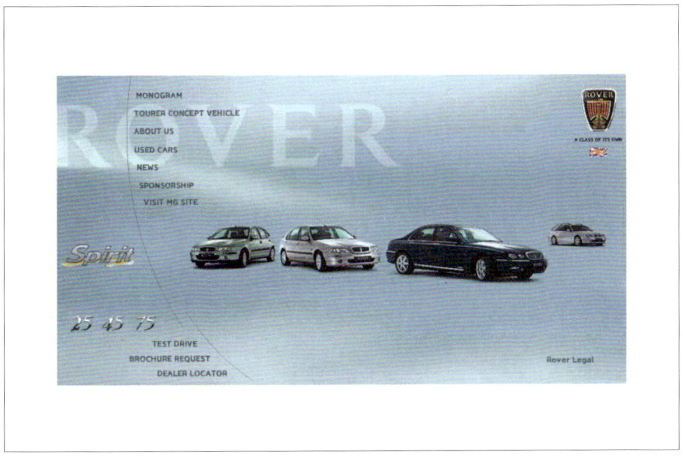

● **ROVER** ● **http://www.mg-rover.com/** ● **UK／2002**
CL, SB：MG Rover Group Limited

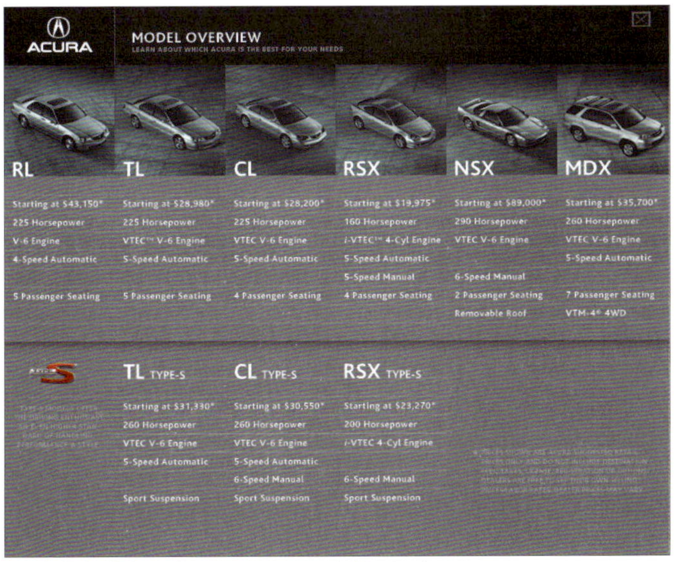

● **ACURA** ● **http://www.acura.com/** ● **USA／2002**
CL, SB：ACURA DIVISION OF AMERICAN HONDA MOTOR COMPANY, INC.

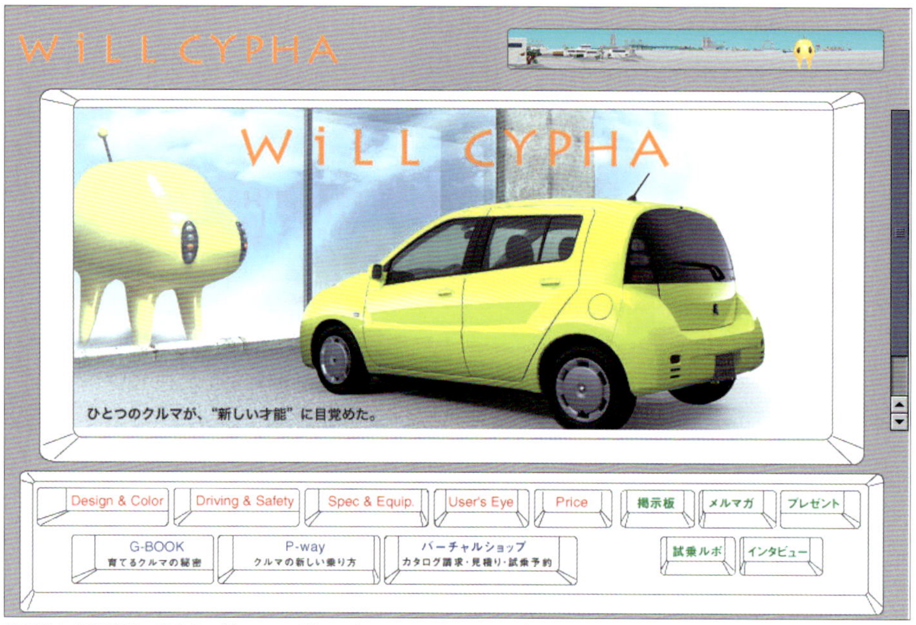

● **WiLL CYPHA** ● **http://www.will-cypha.com** ● **JAPAN／2002**
CL, SB：トヨタ自動車（株） TOYOTA MOTOR CORPORATION DF：（株）博報堂 HAKUHODO Inc.

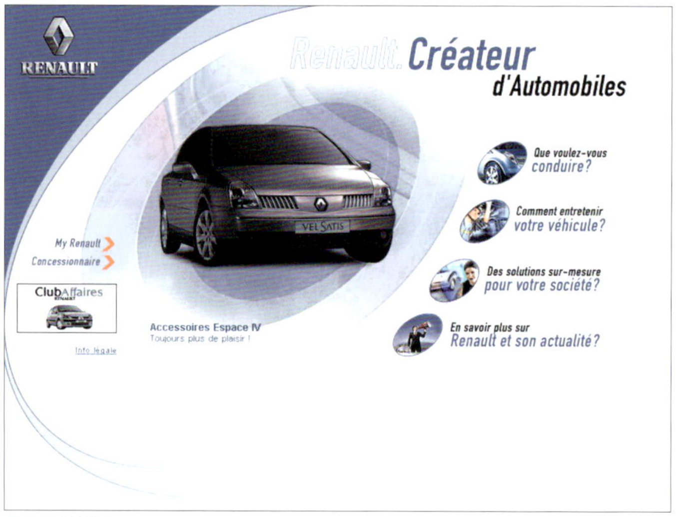

● **RENAULT** ● **http://www.renault.fr/** ● **FRANCE／2002**
CL, SB：RENAULT s.a.s.

● **アピス神戸（APIS COBE）** ● **http://www.apiscobe.co.jp/top.htm**
● **JAPAN／2002**
CL, SB：アピス神戸（株） APIS COBE

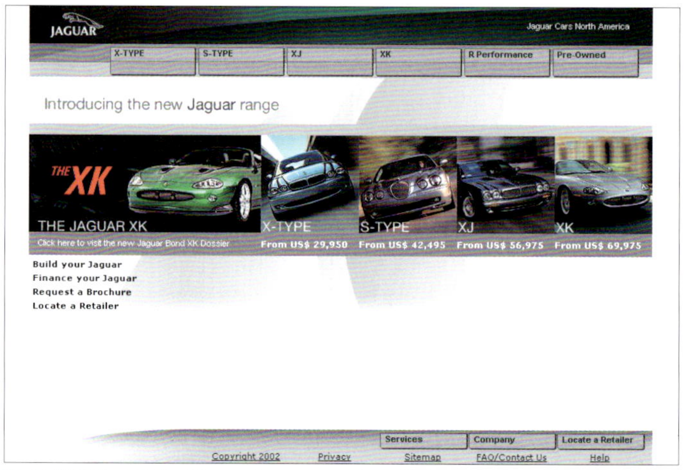

● **JAGUAR** ● **http://www.jaguar.com/** ● **USA / UK／2002**
CL, SB：Jaguar

● **HUSABERG** ● http://www.husaberg.se/ ● SWEDEN／2002
CL, SB：Husaberg Motor AB

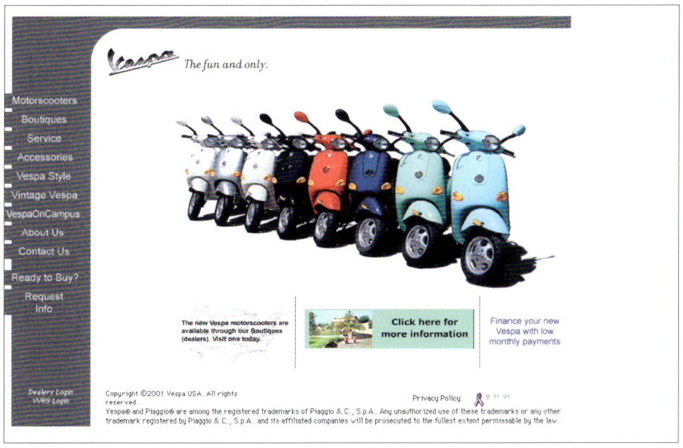

● **Vespa USA** ● http://www.vespausa.com/ ● USA／2002
CL, SB：Vespa USA

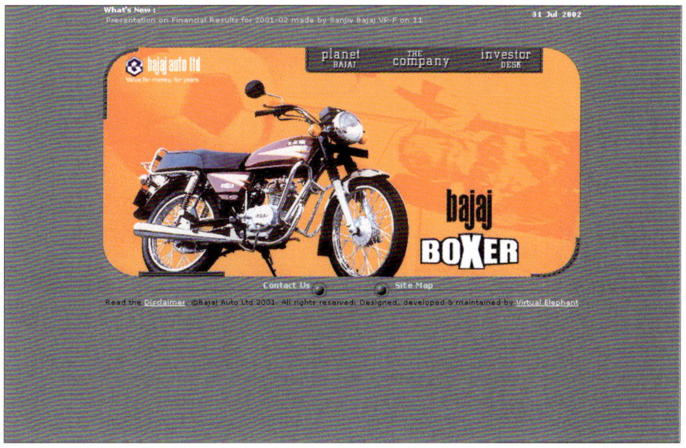

● **bajaj auto** ● http://www.bajajauto.com/ ● INDIA／2002
CL, SB：Bajaj Auto Ltd.

● **PIAGGIO USA** ● http://piaggiousa.com/ ● USA／2002
CL, SB：Piaggio USA

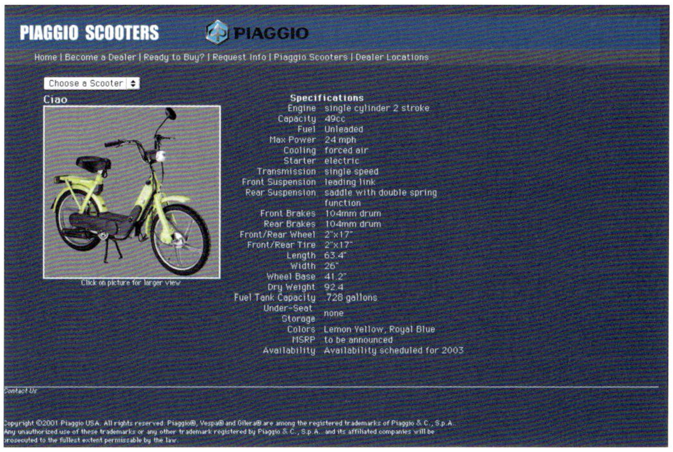

● **Buell** ● **http://www.buell.jp/** ● **JAPAN／2002**

CL, SB：ハーレーダビッドソンジャパン（株） Harley Davidson Japan K.K. DF：（株）レインボージャパン Rainbow Japan

● **YAMAHA**

● **http://www.yamaha-motor.co.jp/mc/index.html** ● **JAPAN／2002**

CL, SB：ヤマハ発動機（株） YAMAHA MOTOR CO., LTD.

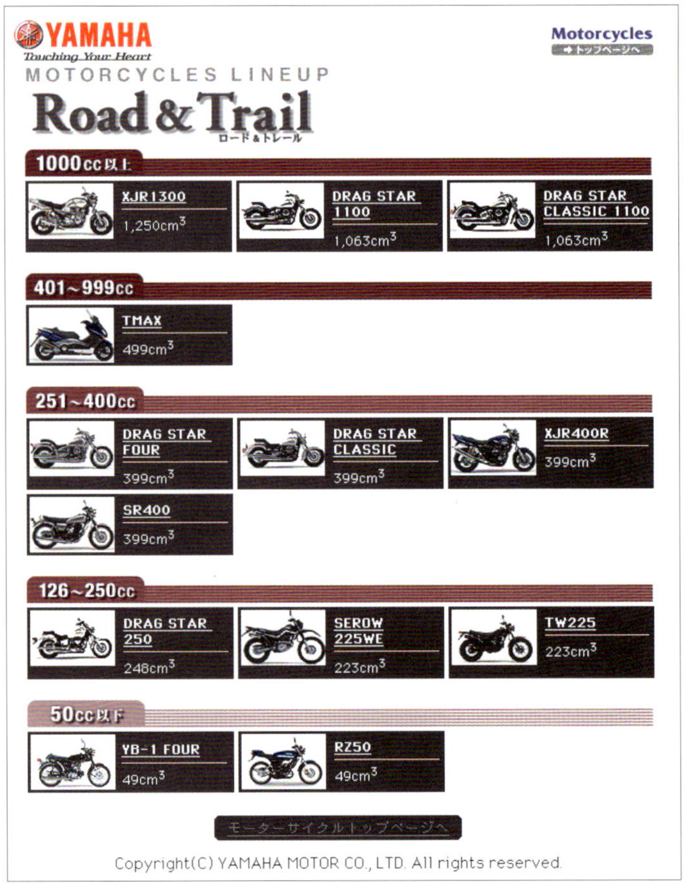

● **YAMAHA**

● **http://www.yamaha-motor.co.jp/mc/street.html** ● **JAPAN／2002**

CL, SB：ヤマハ発動機（株） YAMAHA MOTOR CO., LTD.

● YAMAHA ● http://www.yamaha-motor.co.jp/mc/tmax/ ● JAPAN／2002
CL, SB：ヤマハ発動機（株） YAMAHA MOTOR CO., LTD.

● Indian Motorcycle ● http://www.indianmotorcycle.com/ ● USA／2002
CL, SB：INDIAN MOTORCYCLE CORPORATION

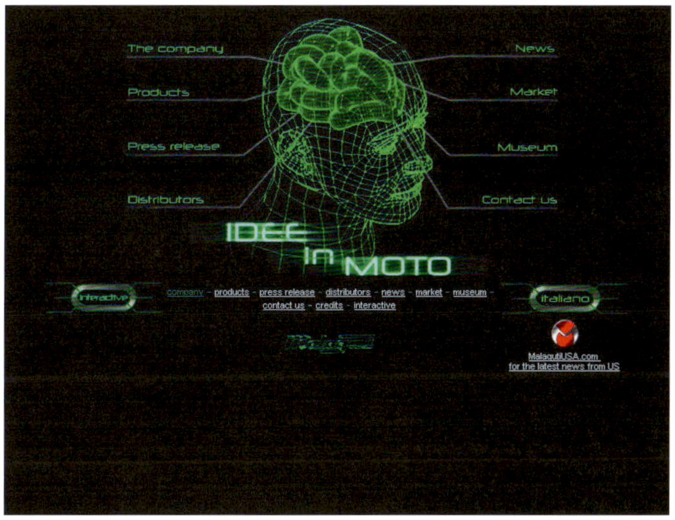

● Malaguti ● http://www.malagutimoto.it/ ● ITALY／2002
CL, SB：Malaguti

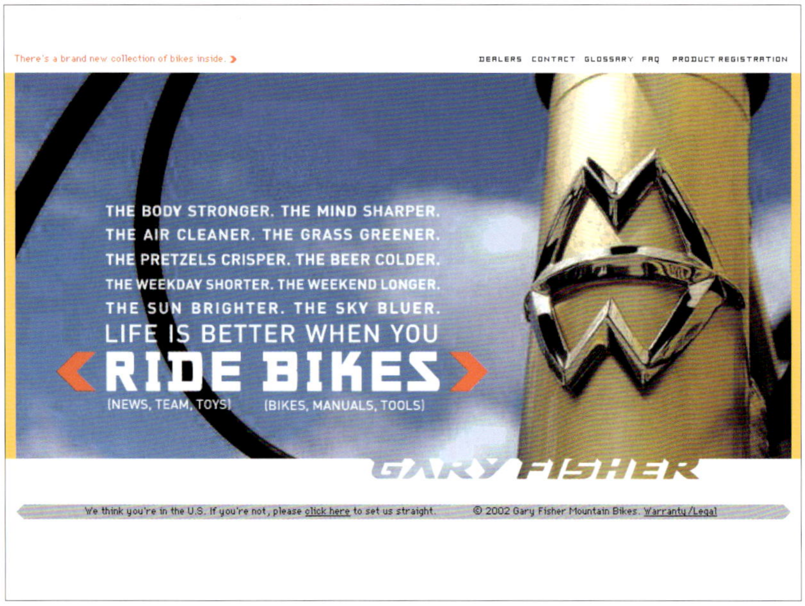

● Gary Fisher bicycles ● http://www.fisherbikes.com
● USA／2002
CL：Trek Bicycles Corp. SB：Planet Propaganda DF：Ben Hirby

concept Gary Fischerマウンテンバイクの公式サイト。伝説的なマウンテンバイクブランドをチューンし、より多くの情報を発信するために作成されました。Gary Fisherの今後の成長の鍵を握るのは、バイクの知識に乏しく、Gary Fischerをよく知らないライダー入門者なので、創設者であるGary Fischer、彼の世界観やバイクを入門者に紹介するようサイトを見直しました。オープンインターフェイスがバイクのガイド役です。「Fisher 101」には、短編のアニメーションフィルムとインタラクティブなスライドショーを採用し、主要なテクノロジーをわかりやすく解説しています。「Gary's Family Room」は、ユーザーが個々の写真、映像、体験談などを共有できる場所です。「Happenings」には最新ニュースとレース結果を掲載。技術的には、Windows2000、ASPおよびSQL Serverを使用してスケーラブル、フレキシブル、高速なサイトを目指しました。バイク検索機能(Matchmaker)はデータ駆動型のFlashアプリケーションで、「Planet」を使用すれば、各モデルイヤーの回答と組み合わされている質問やバイクを変更することができます。クライアント向けに構築したため、更新と管理が簡単にできることを重視しました。

This is the official site for Gary Fisher Mountain Bikes.
Mission: To fine-tune a legendary mountain bike brand and make it accessible to a slightly broader group of consumers.
Method: The majority of growth potential for Fisher lies with "uninitiated" entry-level riders who have less bike knowledge and much less familiarity with Gary Fischer than the core target. We overhauled the site to be able to introduce Gary, his world, and his bikes to this important group, and still preserve the familiar tone in which we speak to our loyal core. Matchmaker uses an open interface to guide users to the bikes just right for them. Fisher 101 has animated short films and interactive slide shows that make key technologies comprehensible. Gary's Family Room is where users share their own photos, movies, and stories. Happenings is where the latest news and up-to-the-minute race results can be found. And there's so much more!
Tech: We used Windows2000, ASP, and SQLServer to build a site that is scalable, flexible, and fast. Matchmaker, the site's bike finder, is a data-driven Flash application that allows Planet to modify the questions and the bikes associated with a particular set of answers for each model year. The administrative Web pages we built for the client make it easy to update the site.

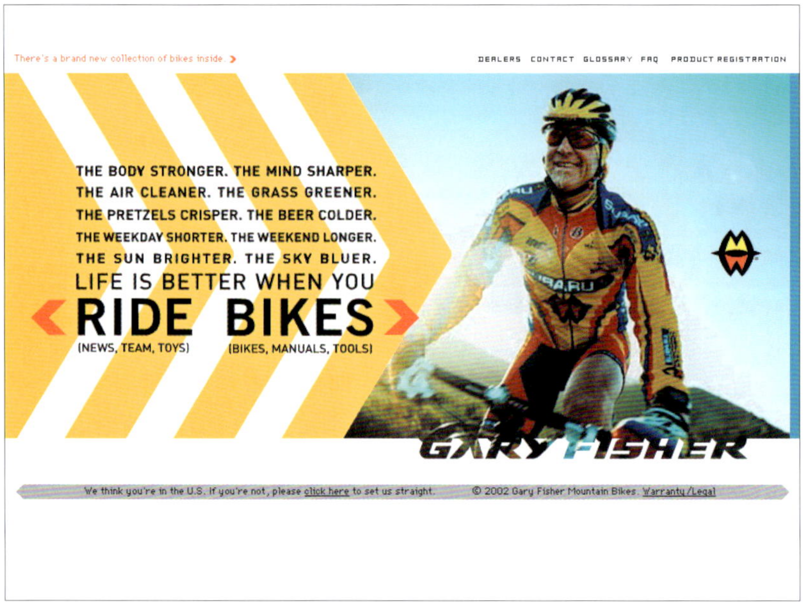

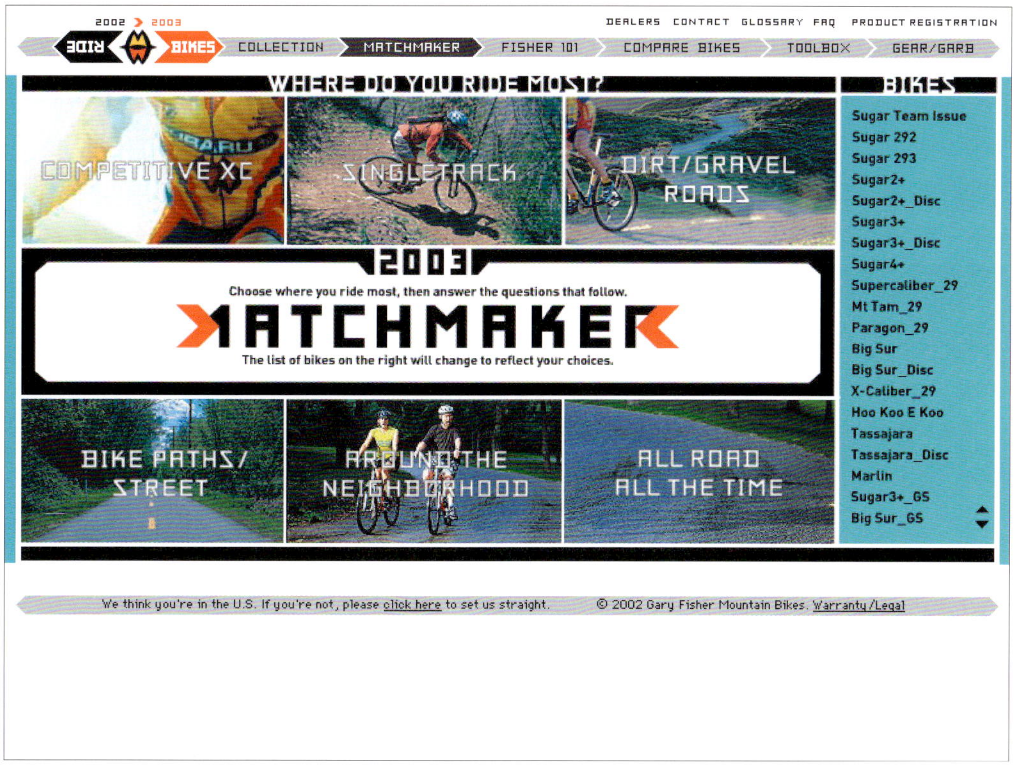
2002 2003
DEALERS CONTACT GLOSSARY FAQ PRODUCT REGISTRATION
RIDE BIKES COLLECTION MATCHMAKER FISHER 101 COMPARE BIKES TOOLBOX GEAR/GARB
WHERE DO YOU RIDE MOST?
COMPETITIVE XC
SINGLETRACK
DIRT/GRAVEL ROADS
2003
Choose where you ride most, then answer the questions that follow.
MATCHMAKER
The list of bikes on the right will change to reflect your choices.
BIKE PATHS/ STREET
AROUND THE NEIGHBORHOOD
ALL ROAD ALL THE TIME
BIKES
Sugar Team Issue
Sugar 292
Sugar 293
Sugar2+
Sugar2+_Disc
Sugar3+
Sugar3+_Disc
Sugar4+
Supercaliber_29
Mt Tam_29
Paragon_29
Big Sur
Big Sur_Disc
X-Caliber_29
Hoo Koo E Koo
Tassajara
Tassajara_Disc
Marlin
Sugar3+_GS
Big Sur_GS
We think you're in the U.S. If you're not, please click here to set us straight.
© 2002 Gary Fisher Mountain Bikes. Warranty/Legal

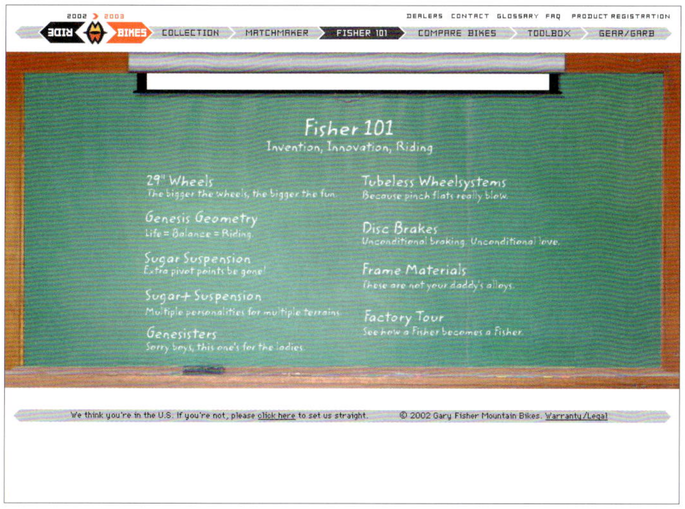
2002 2003
DEALERS CONTACT GLOSSARY FAQ PRODUCT REGISTRATION
RIDE BIKES COLLECTION MATCHMAKER FISHER 101 COMPARE BIKES TOOLBOX GEAR/GARB
Fisher 101
Invention, Innovation, Riding
29" Wheels
The bigger the wheels, the bigger the fun.
Genesis Geometry
Life = Balance = Riding.
Sugar Suspension
Extra pivot points be gone!
Sugar+ Suspension
Multiple personalities for multiple terrains
Genesisters
Sorry boys, this one's for the ladies.
Tubeless Wheelsystems
Because pinch flats really blow.
Disc Brakes
Unconditional braking. Unconditional love.
Frame Materials
These are not your daddy's alloys.
Factory Tour
See how a Fisher becomes a Fisher.
We think you're in the U.S. If you're not, please click here to set us straight.
© 2002 Gary Fisher Mountain Bikes. Warranty/Legal

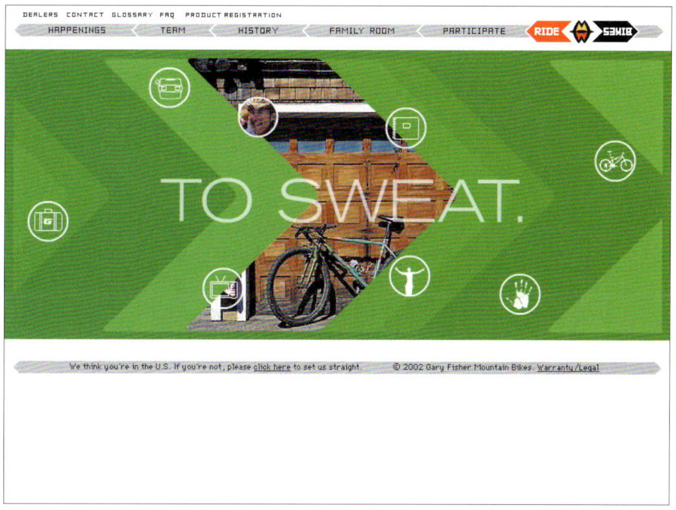
DEALERS CONTACT GLOSSARY FAQ PRODUCT REGISTRATION
HAPPENINGS TEAM HISTORY FAMILY ROOM PARTICIPATE RIDE BIKES
TO SWEAT.
We think you're in the U.S. If you're not, please click here to set us straight.
© 2002 Gary Fisher Mountain Bikes. Warranty/Legal

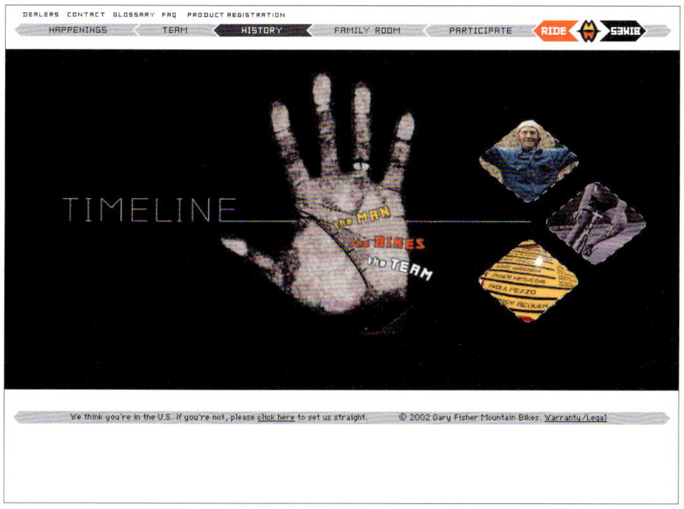
DEALERS CONTACT GLOSSARY FAQ PRODUCT REGISTRATION
HAPPENINGS TEAM HISTORY FAMILY ROOM PARTICIPATE RIDE BIKES
TIMELINE
the MAN
the BIKES
the TEAM
We think you're in the U.S. If you're not, please click here to set us straight.
© 2002 Gary Fisher Mountain Bikes. Warranty/Legal

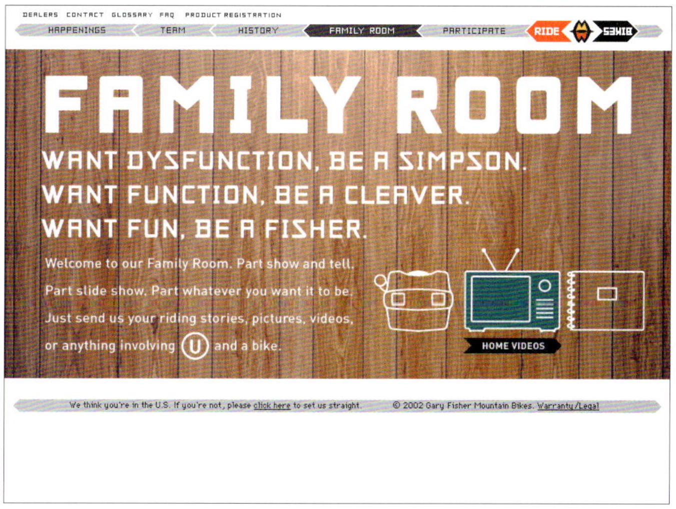
DEALERS CONTACT GLOSSARY FAQ PRODUCT REGISTRATION
HAPPENINGS TEAM HISTORY FAMILY ROOM PARTICIPATE RIDE BIKES
FAMILY ROOM
WANT DYSFUNCTION, BE A SIMPSON.
WANT FUNCTION, BE A CLEAVER.
WANT FUN, BE A FISHER.
Welcome to our Family Room. Part show and tell.
Part slide show. Part whatever you want it to be.
Just send us your riding stories, pictures, videos,
or anything involving U and a bike.
HOME VIDEOS
We think you're in the U.S. If you're not, please click here to set us straight.
© 2002 Gary Fisher Mountain Bikes. Warranty/Legal

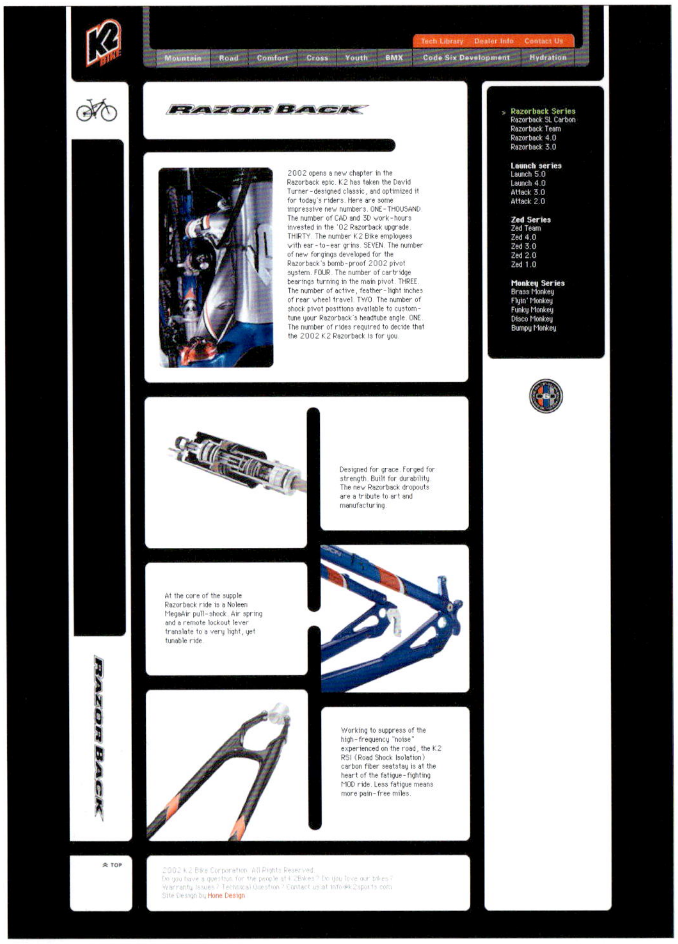

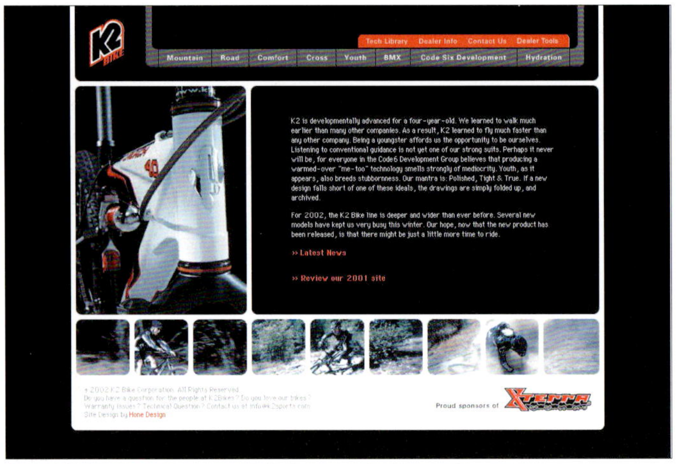

● K2 BIKE ● http://www.k2bikes.com/ ● USA／2002
CL, SB： K2 Bike Corporation.

● Pinarello.com ● http://www.pinarello.com/ ● ITALY／2002
CL, SB： Cicli PINARELLO S.p.A.

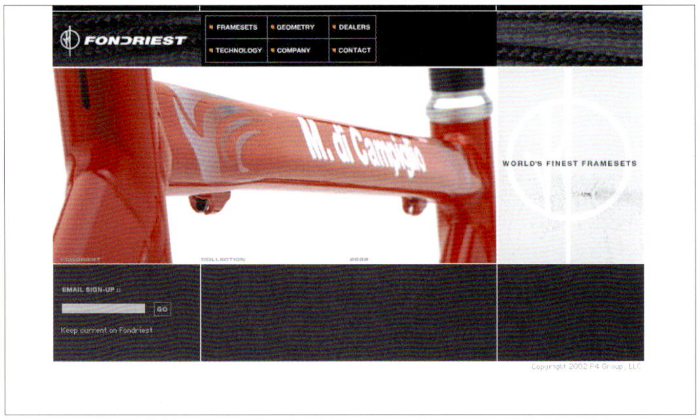

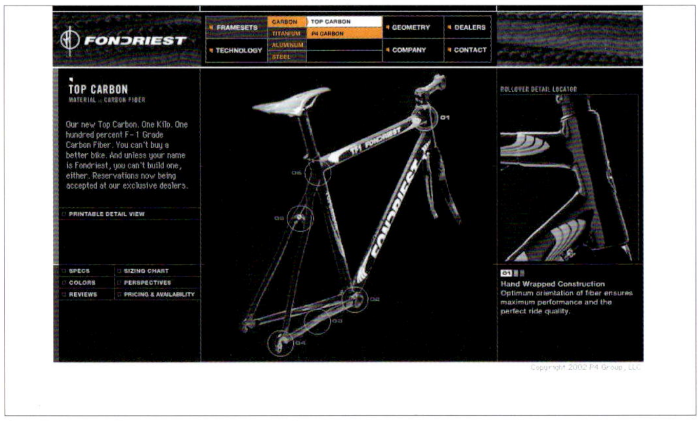

● FONDRIEST ● http://www.fondriest-usa.com/
● USA／2002
CL, SB： P4 Group, LLC.

● LEE COUGAN ● http://www.leecougan.com/ ● USA／2002
CL, SB：Lee Cougan

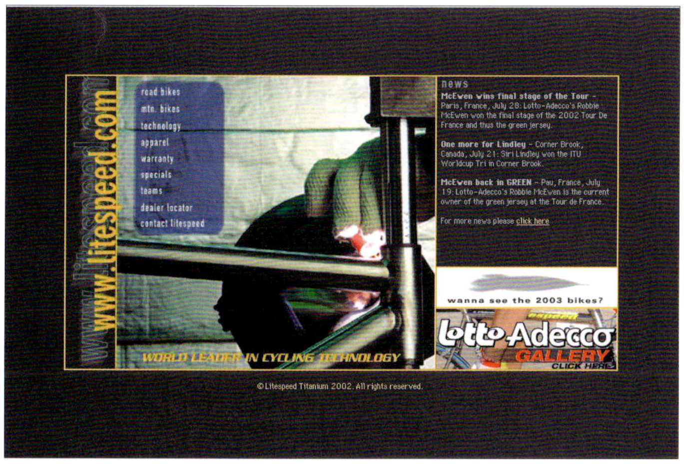

● Litespeed Titanium ● http://www.litespeed.com/ ● USA／2002
CL, SB：Litespeed Titanium

● HOFFMAN ● http://www.hoffmanbikes.com/ ● USA／2002
CL, SB：Hoffman Bikes

● YETI CYCLES ● http://www.yeticycles.com/ ● USA／2002
CL, SB：Yeti Cycles

● MERLIN ● http://www.merlinbike.com/ ● USA／2002
CL, SB：Merlin Metalworks

● SCHWINN ● http://www.schwinnbike.com/ ● USA／2002
CL, SB：Pacific Cycle, LLC.

● HUFFY ● http://www.huffybikes.com/ ● USA／2002
CL, SB：Huffy Bicycle Company

● SCAPIN ● http://www.scapin.com/ ● ITALY／2002
CL, SB：Scapin Bike S.r.l.

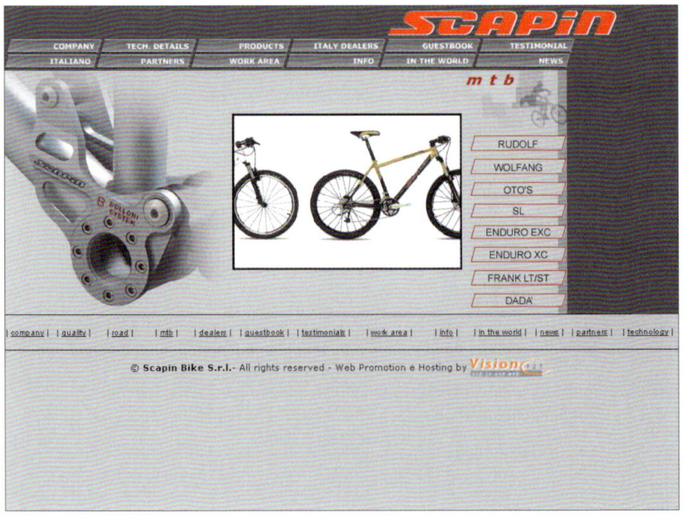

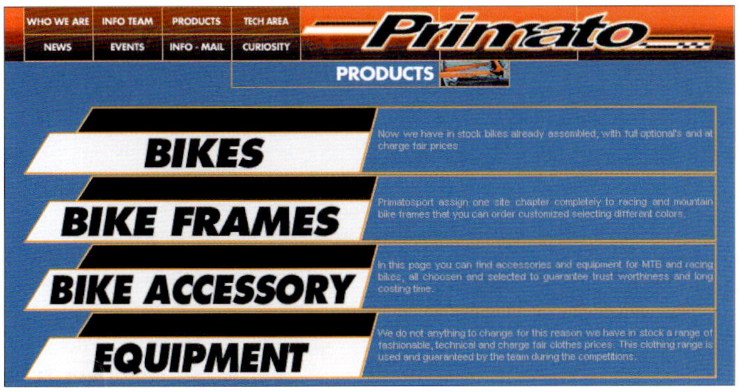

● Primato ● http://www.primatosport.com/ ● ITALY／2002
CL, SB： Primatosport

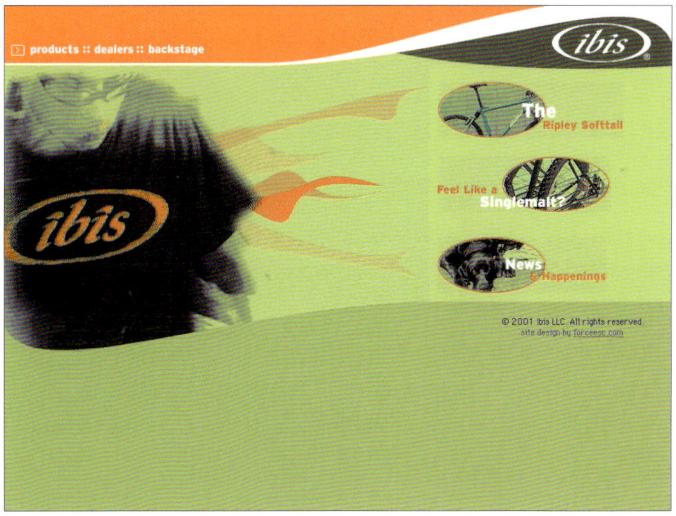

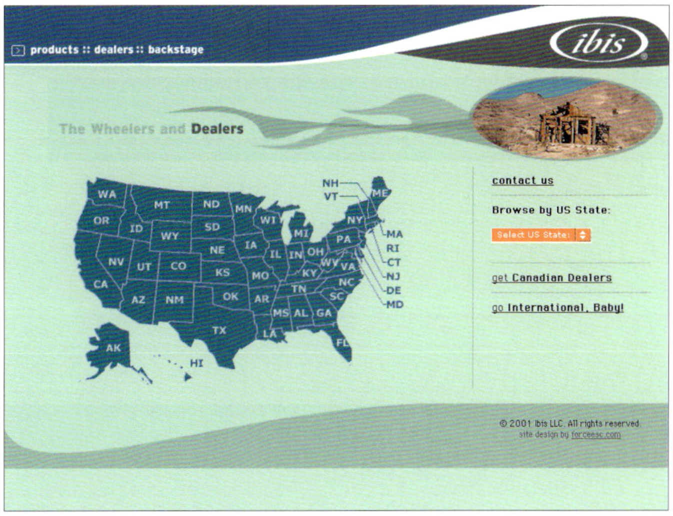

● Ibis ● http://www.ibiscycles.com/ ● USA／2002
CL, SD： Ibis LLC.

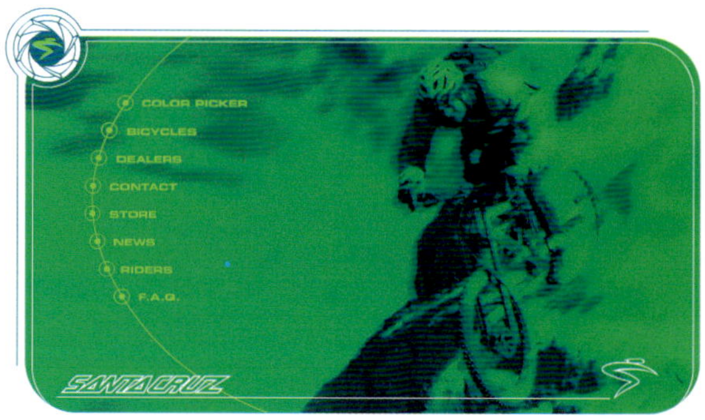

● **SANTA CRUZ** ● **http://www.santacruzmtb.com/** ● **USA／2002**
CL, SB：Santa Cruz Bicycles

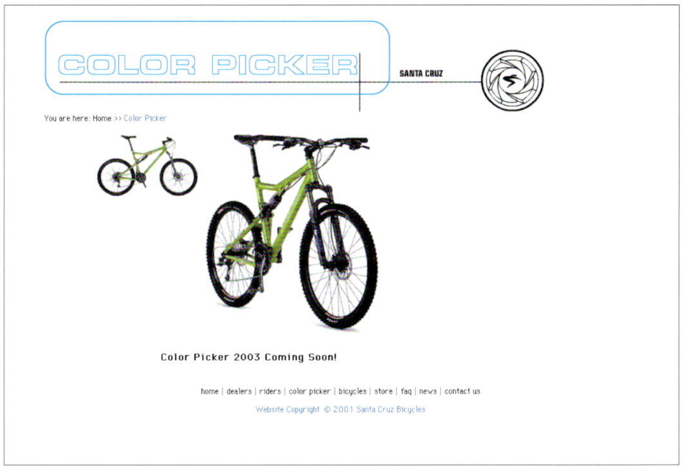
COLOR PICKER
SANTA CRUZ
You are here: Home >> Color Picker
Color Picker 2003 Coming Soon!
home | dealers | riders | color picker | bicycles | store | faq | news | contact us
Website Copyright © 2001 Santa Cruz Bicycles

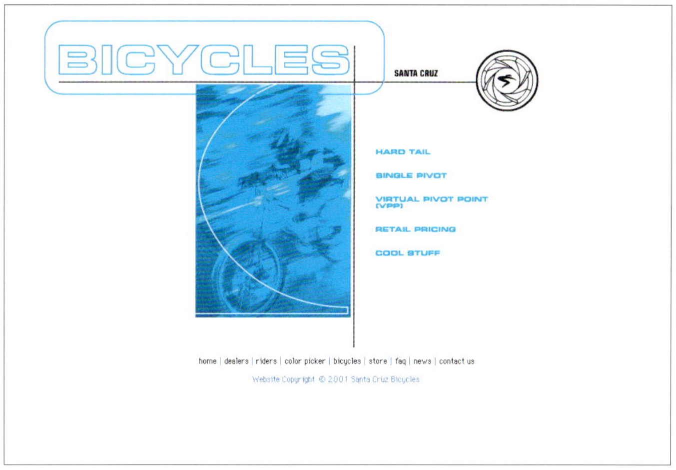
BICYCLES
SANTA CRUZ
HARD TAIL
SINGLE PIVOT
VIRTUAL PIVOT POINT (VPP)
RETAIL PRICING
COOL STUFF
home | dealers | riders | color picker | bicycles | store | faq | news | contact us
Website Copyright © 2001 Santa Cruz Bicycles

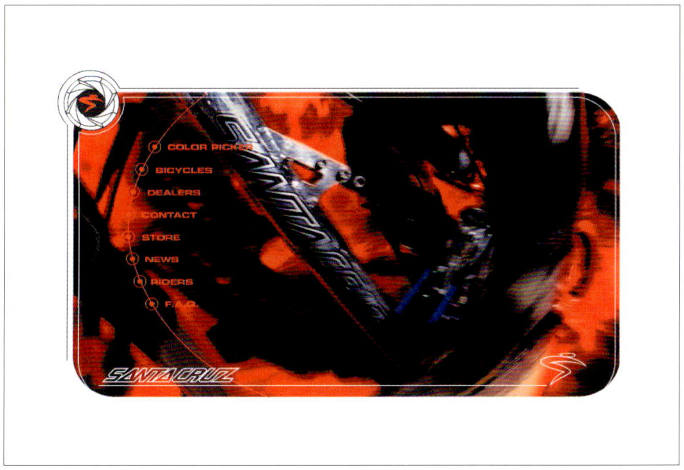
COLOR PICKER
BICYCLES
DEALERS
CONTACT
STORE
NEWS
RIDERS
F.A.Q.
SANTA CRUZ

COLOR PICKER
DEALERS
CONTACT
SANTA CRUZ

COLOR PICKER
BICYCLES
DEALERS
CONTACT
STORE
NEWS
F.A.Q.
SANTA CRUZ

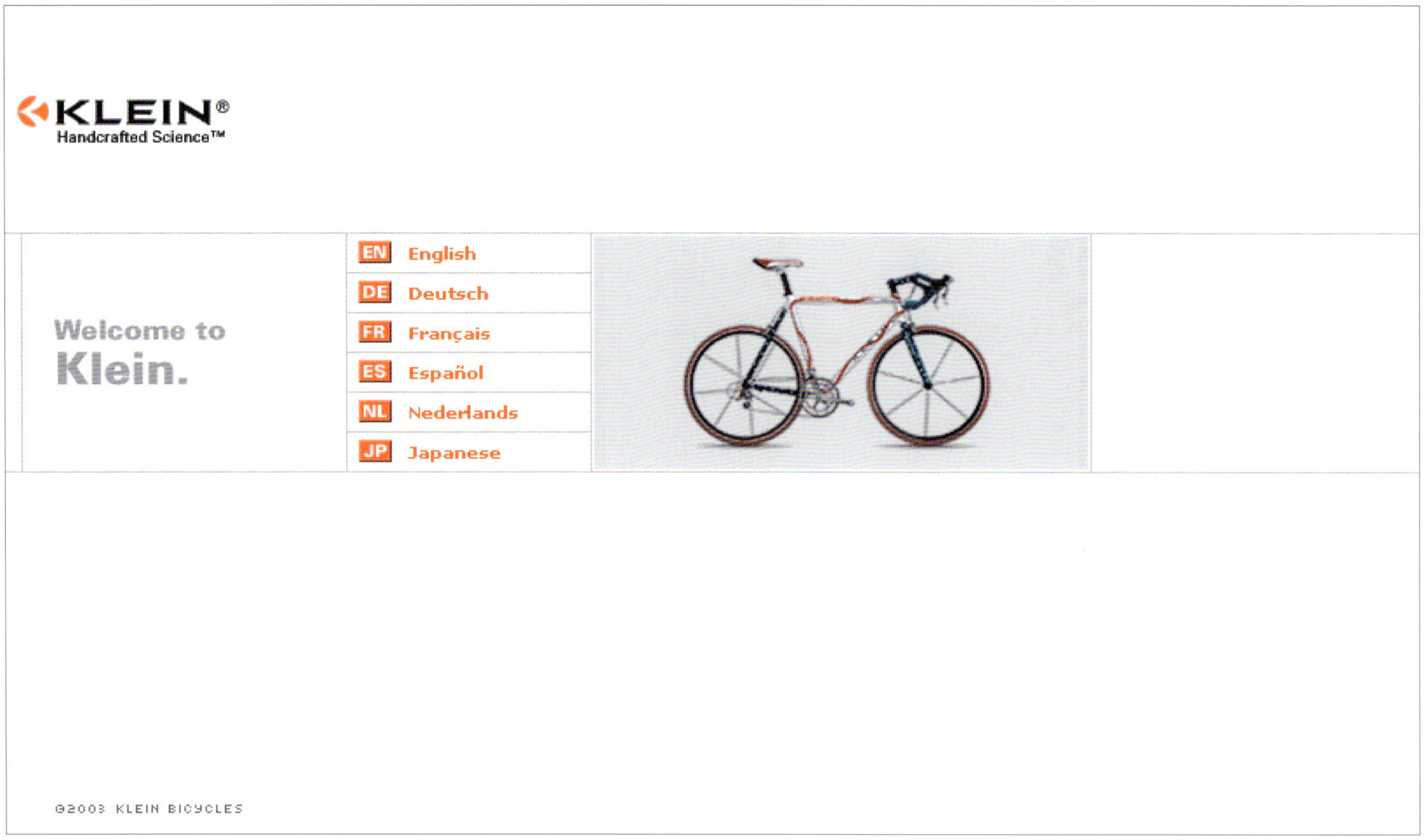

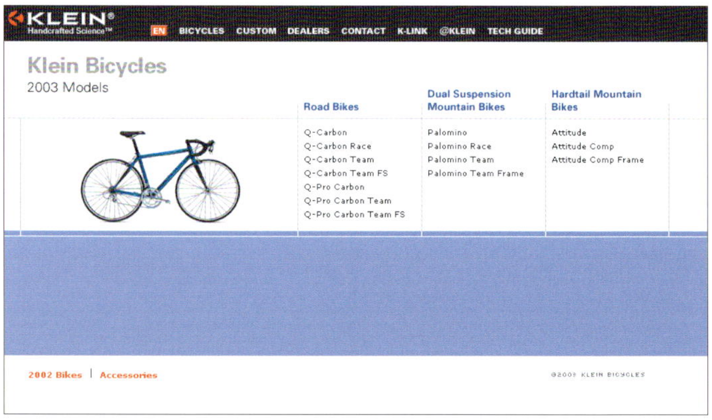

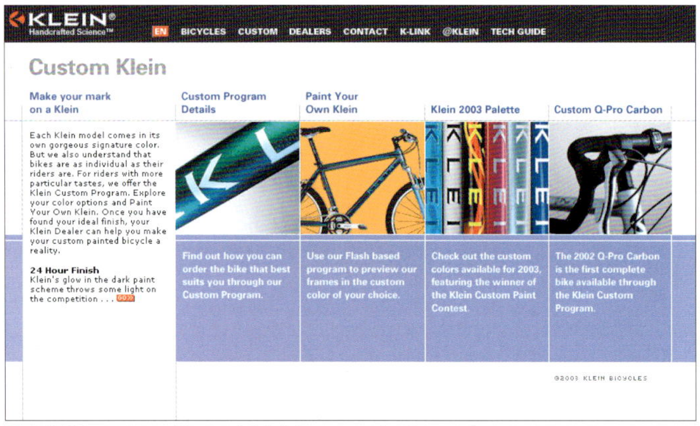

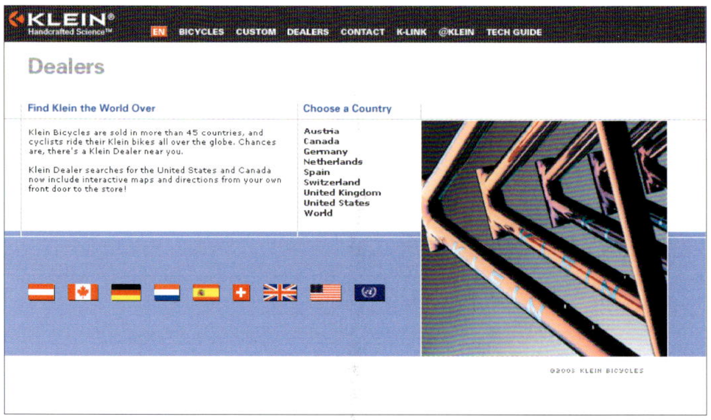

● **KLEIN** ● **http://www.kleinbikes.com/** ● **USA／2002**
CL, SB：KLEIN BICYCLES

料金受取人払

豊島局承認

6265

差出有効期間
平成17年3月
4日まで

郵便はがき

1 7 0 - 8790

038

東京都豊島区駒込4-14-6-301

ピエ・ブックス 行

愛読者カード

このたびは小社の本をお買い上げいただき、ありがとうございます。新刊案内の送付と、今後の企画の参考とさせていただきますので、お手数ですが各欄にご記入の上ご郵送下さい。

書名　**業種別 ウェブメニュー コレクション**

1. 本書についてのご感想をお聞かせ下さい。また、今後の出版物についてどのようなテーマを望まれますか。

ご感想

望むテーマ

2. 本書をお買い上げいただいた動機。

A. 広告で（紙・誌名　　　　　　）　B. 知人にすすめられて
C. 書評で（紙・誌名　　　　　　）　D. 書店で見て
E. 小社刊行物で　F. DMで　G. その他（　　　　　　）

3. いままで読者カードをお出しいただいたことが

A. ある　　B. ない

ふりがな
お名前　　ご職業　（　　才）

会社名

ご住所（〒　　　　）

ビル名

TEL.　　FAX.

e-mail

ご購入書店名

アンケートにご協力いただきありがとうございました。

OTHERS

Artist & Creater
［アーティスト＆クリエイター］

Production
［制作会社］

Museum & Gallery
［美術館、博物館＆ギャラリー］

Others
［その他］

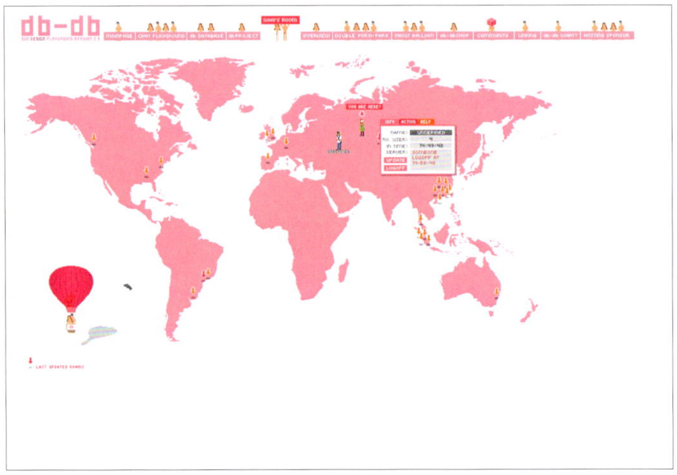

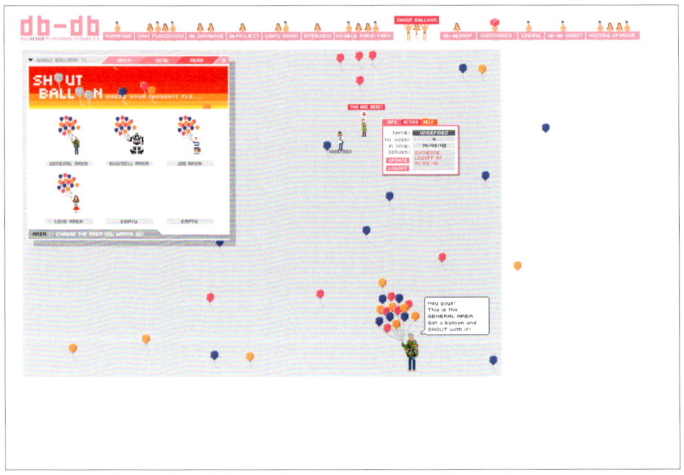

● **db-db** ● **http://db-db.com** ● **CHINA／2002**

CL, DF：Francis Lam SB：db-db

concept 「db-db」はデザインとインスピレーションの遊び場です。「db」の文字の形が気に入ってこの名前をつけました。大きな鼻が2つ並んでいるみたいでかわいいでしょう？

db-db is my playground of design and inspiration. I named the site "db" simply because the letters look like a big nose, and two big noses look more lovely!

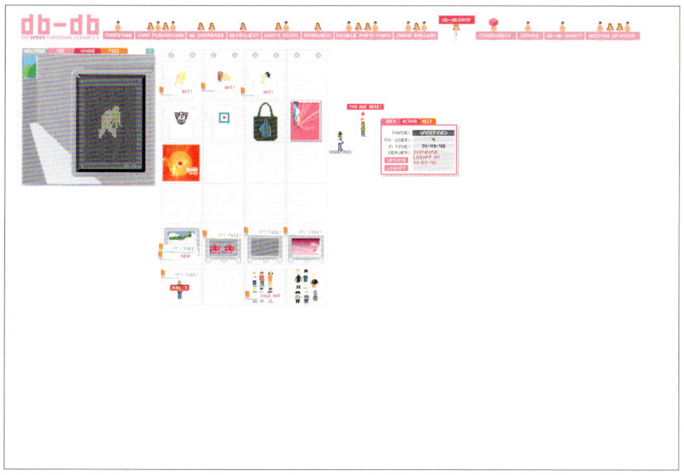

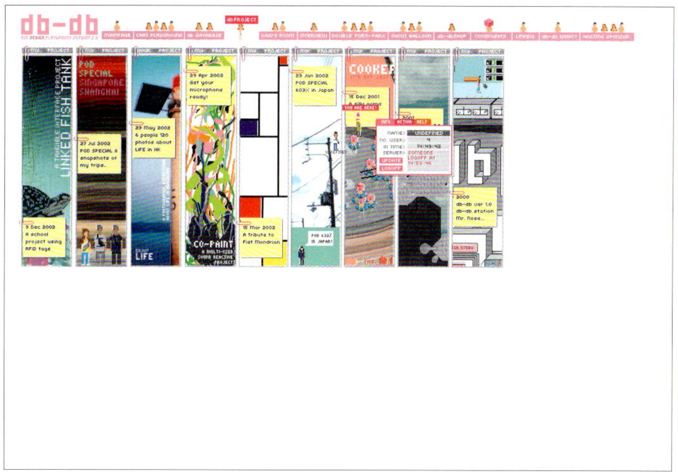

db-db
CHAT PLAYGROUND
MAINPAGE
db DATABASE
db PROJECT
WHO'S ROOM
INTERVIEW
DOUBLE PORN-PARK
SHOUT BALLOON
db-dbSHOP
COMMUNITY
LINKING
db-db WHAT?
HOSTING SPONSOR
YOU ARE HERE!
INFO
ACTION
HELP
NAME>
UNDEFINED
NO. USER>
4
IN TIME>
14:49:42
SERVER>
SOMEONE
LOGOFF AT
14:58:46
UPDATE
LOGOFF
UNDEFINED
DB-DB AIR FORCE
923-78

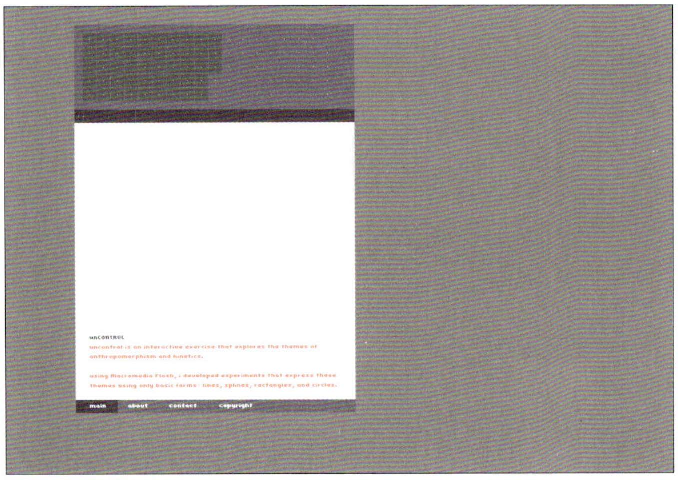

● **Uncontrol** ● **http://www.uncontrol.com** ● **USA／2002**
CL, DF：Manny Tan　SB：Uncontrol

concept これまで、動作や機能を豊富に備えたWebサイトを数多くプログラミングしてきましたが、読みやすさとスペースを損なわずにすべての要素を表示することが大きな課題でした。作者が決めたルートをたどらなくてもサイト内を探索できるように、旧バージョンのモチーフであったグリッドとボックスのみによる単純な構成に仕上げげました。以前訪問したことのあるWebサイトを再度訪問したとき、前にどのコンテンツをクリックしたのか忘れてしまった経験はありませんか？　閲覧したコンテンツを確認するには、マークを付けるだけでは不充分です。そこで閲覧したルートをたどるラインを採り入れました。コンテンツをクリックするほど数多くのラインが網状にはりめぐらされ、もっとクリックしたいという気持ちをあおる仕組みです。機能的には最近のWebサイトによく見られるパンくずリストと同じですが、パンくずリストのようにルートを文字で表示するだけでなく、クリックのルートをラインで描き、さらにその下にラベルを表示するよう工夫しました。クリックすると、コンテンツを示すラベルが右に追加されていき、最初にクリックしたコンテンツから現在表示しているコンテンツまでの履歴を簡単に把握することができます。履歴バーには5回前の訪問までさかのぼって表示されますが、スクロールバーを使えばそれ以上の履歴を表示することもできます。このレイアウトは当初次のデザインを考案するまでの応急策でしたが、予期していた以上の素敵な仕上がりになりました。

Over the past few years I have created a variety of personal experiments involving programmatic movement and behavior. I needed a mechanism with which to display all the pieces without sacrificing readability and space. I want people to be able explore the experiments without having to stick to a particular route, so I decided to keep the motif of simple grids and boxes I used in prior versions. I have a short attention span and usually don't remember what I clicked last on a particular page, so just marking the experiments wasn't enough. I want visitors to see where they have been, so I drew lines from the previous experiment to the current experiment. As visitors click on different experiments, more lines are generated and create a web-like form that encourages visitors to keep clicking. This is a visual representation of the bread-crumbing feature found in most sites—an idea I wanted to expand on. I included a linear bread-crumbing format below the grid that doubles as a label for the system; with each click it adds another word to the right, building a readable history of where you started to where you currently are. Since the history bar can continue indefinitely, I allotted space on the screen to display the most recent five experiments; the user has the option of using the scrollbars to see more. Interestingly enough, this layout was intended as a stopgap for my next design, but it I liked it so much I kept it.

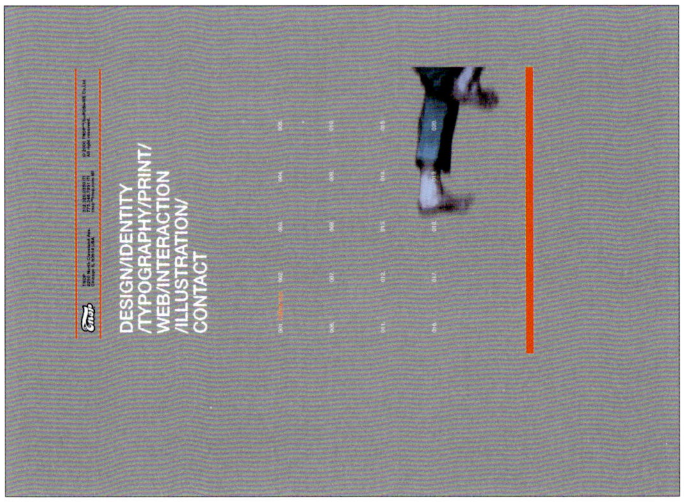

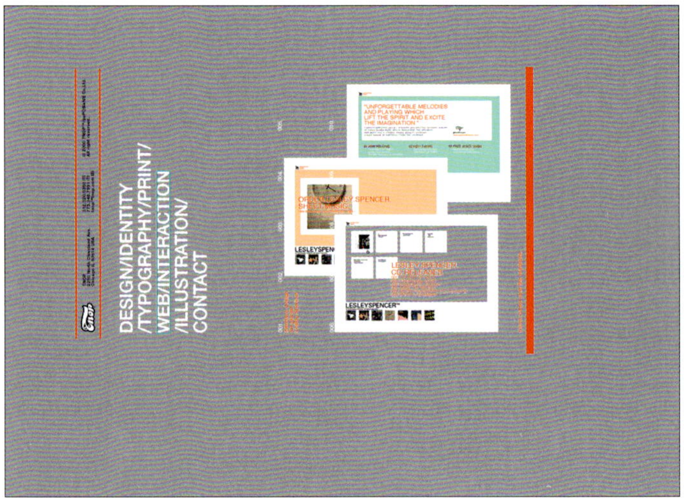

● **Tnop** ● **http://www.tnop.com** ● **USA／2000**
CL：Tnop™　SB：®bePOS|+|VE design(www.bepositivedesign.com)　DF：Tnop Wangsillapakun

concept 技術革新の早さとデジタルメディアに適用されるリソースには驚かずにいられません。Webデザイナーはしのぎ合いに懸命で、最新ソフトウェアを使って奇抜なコンテンツを作成することに躍起となっています。その結果、芸術的には決して高品質と言えない複雑なだけのデザインが蔓延しています。そこで、マウスを利用するシンプルなナビゲーションシステムを採用しました。番号の上にカーソルを重ねるだけで作品が表示されます。

I am amazed at the rapidity of technological advancement and the incredible resources that are being applied to digital media. Web designers scramble to outdo each other, creating increasingly fantastic effects with the latest software. The result is complicated design that lacks artistic quality. I decided to keep my Web portfolio simple by using a "roll over" navigation technique. All visitors have to do is roll the Mouse over the numbers to display my work.

● **State** ● **http://www.statedesign.com**
● **UK／2001**
CL, SB：state DF：Philip O' Dwyer

concept さまざまなページをまとめた総合的なWebサイト。グリッドで仕切られたページにstateの全作品が整然と並べられています。今後もプロジェクトを終えるごとにコンテンツが増えていきます。サイト内を閲覧すると、そのルートをたどるラインが描かれます。

The conventional Web site splits up into separate pages and areas. We wanted to present a picture of state all on one sheet, with a grid of projects that fills up over time. As users travel around the grid, they leave trails showing where they have been.

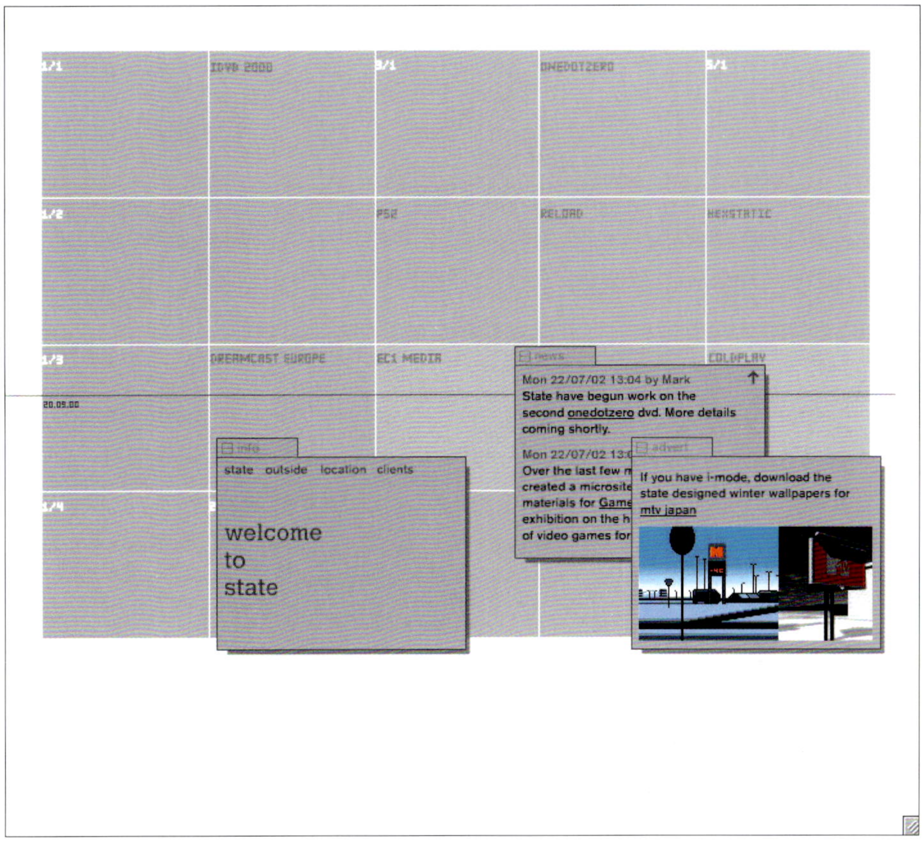

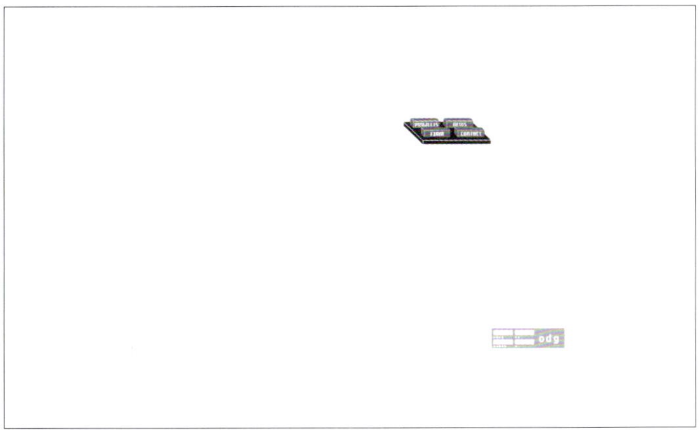

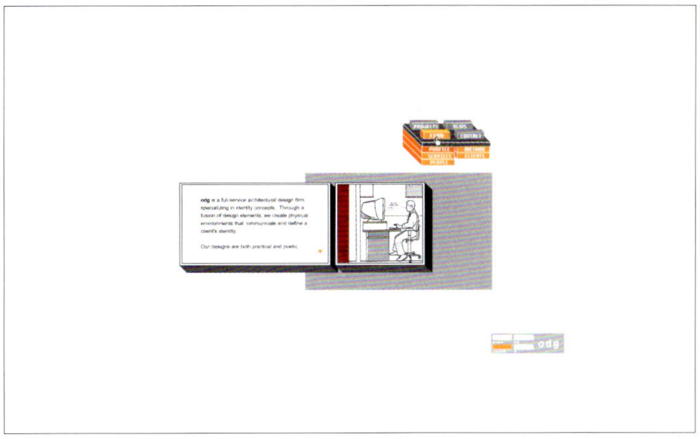

● **odg/otto design group** ● **http://www.ottodesigngroup.com** ● **USA／2001**
CL, SB：odg/otto design group DF：160 over 90 with odg

concept ODGの社長であるPhilip Ottoは「一般に、顧客のビジネスはデザインとイメージ次第で決まります」と述べています。「クライアントのブランドとアイデンティティを的確にとらえて物理的な領域に変換する能力を実証するには、Webサイトが必要でした。広告代理店を何社もわたり歩いた結果、際立っていたのが160over90です。目を見張る創作センスを実証しただけでなく、我々のビジネスとWebサイトにおける達成目標を的確に理解してくれました」。
「建築は広告と同様、ブランドとアイデンティティを物理的領域に変換する媒体なのです」と、160over90のエグゼクティブディレクティブディレクター（executive directive director）のDarryl Cilliは述べています。「ODGは、ビルやインテリアを広告媒体であり、また顧客がブランドを構築するためのきっかけを与えるものと考えています。Webサイトは、このコンセプトを反映させる真っ白いキャンバスです。その結果がODGの清潔でシンプル、かつ芸術的なサイトにあらわれています。

"Our typical client's business hinges on design and image," said Philip Otto, president of odg. "We needed to create a site that demonstrates our ability to capture the client's brand and identity and translate it to a physical space. We looked at a lot of ad agencies; 160over90 stood out, because they demonstrated an amazing creative sense and were able to understand our business and what we needed to achieve with this Web site from the onset."
The site focuses on communicating the breadth and depth of the high-profile work odg has done for clients around the world in retail, office, and residential design.
"Architecture, like advertising, is a medium to translate one's brand and identity, but in a physical space," said Darryl Cilli, executive creative director at 160over90. "Odg looks at buildings and their interiors as a form of advertising and another opportunity for its clients to create a relevant brand experience. We viewed the Web site as a blank canvas on which to apply this concept. The result is a site design that reflects odg's clean, simple, and direct aesthetic."

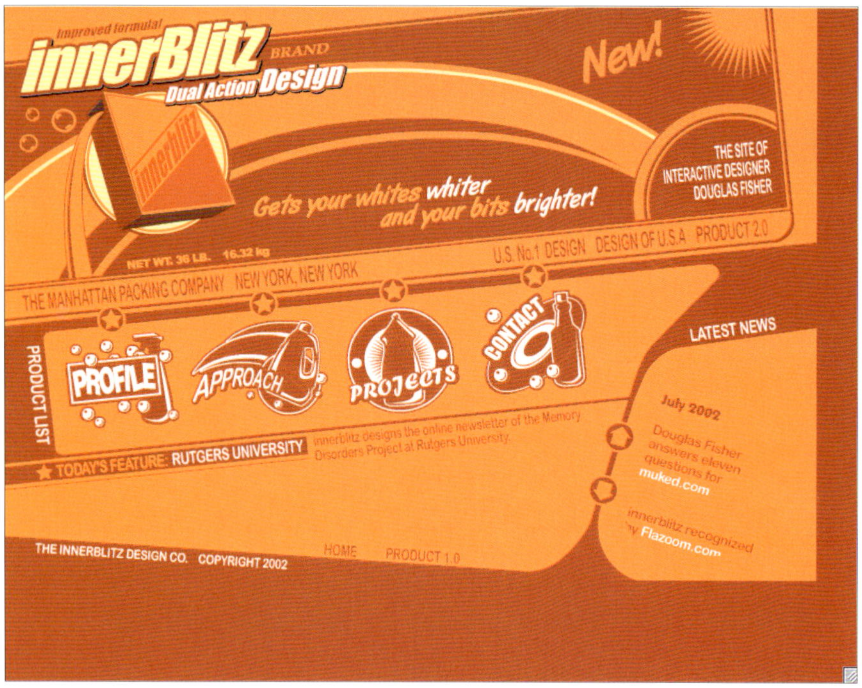

● **innerBlitz** ● **http://www.innerblitz.com**
● USA／2002
CL, SB：innerblitz DF：Douglas Fisher

concept マンハッタン・アッパーイーストのスーパーマーケット「Gristede's」の7番レーンで生まれたinnerblitzのWebサイト。コンテンツをご覧になれば、ただの「技術屋」ではないことがおわかりになるはずです。インタラクティブデザインを決定づけるのは幻想的なアニメーションや派手な色使いのグラフィックだけではありません。innerblitzは、ビジュアルのデザインと開発を通した効果的な伝達と表現に注目しています。電子媒体がユーザーの強力なツールとしても企業が戦略的成長を遂げるための情報伝達手段としても重要であることを強調しています。

Innerblitz was born in aisle 7 of the local Gristede's supermarket on the Upper East Side of Manhattan. As you can see, it was not the "techno" aisle. To understand interactive design is to understand that it exists beyond just fancy animations and big colorful graphics. Innerblitz focuses on effective communication and presentation through visual design and development. It stresses the importance of the electronic medium as both a powerful end-user tool and a vehicle for the strategic growth of a company.

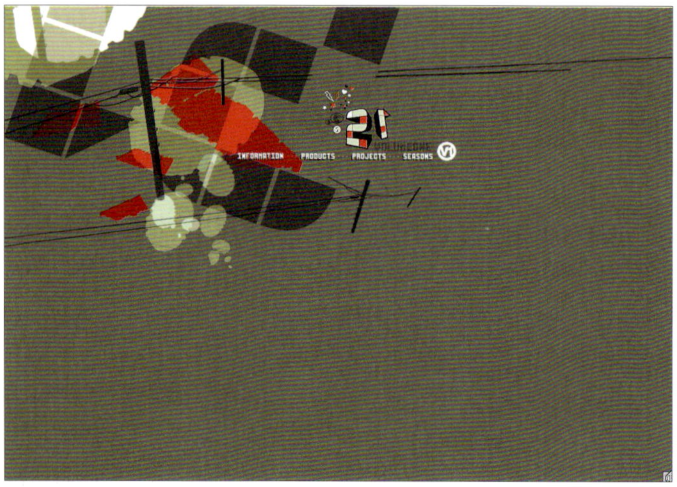

● **volumeone 21** ● **http://www.volumeone.com** ● USA／2002
CL, DF：Matt Owens SB：volumeone

concept 自分の作品の展示スペースとして立ち上げました。就職後、職業や住まいを転々としながら生活していた私がやっと落ち着いたとき、自分のコレクションを展示する場が欲しくなりました。学生時代にWebデザインの経験があったためにWebの活用を思いついたのですが、学んできたとはいえ、Webに精通しているわけでもなければ、Web言語を操る能力もありません。単に好きな作品をほかの人と共有したいというよくある発想です。このWebサイトを通じてこれまでの作品を振り返り、学習と実験に活用したいと考えています。愚作も自分なりの傑作もありますが、自分を駆り立てる仕事の本質を確認することができるのです。

I made volumeone because I wanted to start doing my own stuff again. I got out of school and started working, moved, and had to make money to survive. After I got settled, I felt like I needed an outlet to keep doing work I enjoy. I did some Web design in school that seemed like a good idea at the time. I am not a total Web person/coder, although I have learned a lot of that over the years. I just want to make work I like and share it with others—good or bad, just like any other zine. I look back on the work I have done and see it as a small part of a long process of learning and experimenting. Some of it is lame and some of it I still like, but being able to see the continuum and body of work is what interests me.

● **DHKY** ● **http://www.dhky.com** ● USA／1997
CL, SB：DHKY DF：David Yu

concept 作者とそのクライアントのオンライン作品集として1997年に開設され、以後、デザインとインタラクティビティを模索する試験的なプラットフォームへと変化しました。ここでは、デザインを練習したり、他では採用されないような新しいアイデアを自由に試したりすることができます。感覚に訴え、遊び心に富み、楽しくて使いやすいインターフェイスやメニューを、デザイン全体に適用させる工夫をしました。作品のテーマはアジアやアジア系アメリカのポップカルチャーが中心ですが、それらの枠にとらわれることなく活動していきたいと考えています。

DHKY started up in 1997 as an online portfolio of my client and personal work. It later became an experimental platform for exploring design and interactivity. The site gives me an excuse to practice design and freedom to play around with new ideas that may otherwise not be actualized. I am particularly interested in interfaces and menu systems that are well integrated into the overall design, tactile and responsive, and are playful, fun, and easy to use. Thematically, much of my work centers around Asian and Asian-American pop culture but is not necessarily bound by it.

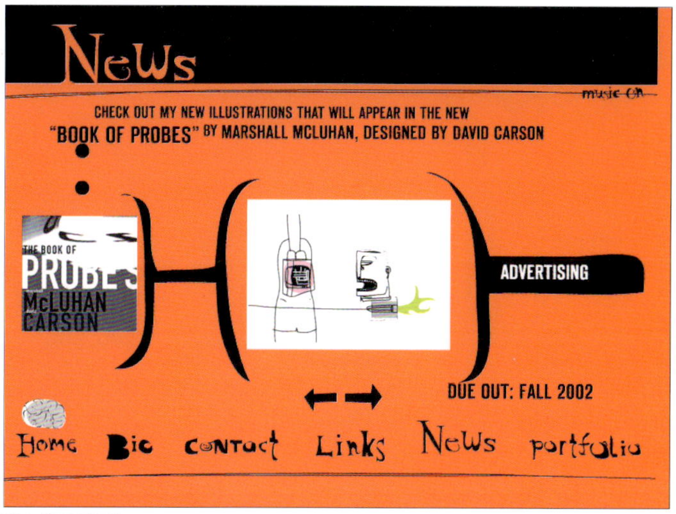

● **MICHAEL LOTENERO ILLUSTRATION + DESIGN** ● **http://www.lotenero.com** ● USA／2001/2002

CL, SB：Michael Lotenero Illustration + Design DF：Michael Lotenero /Scott Bowlin

concept 楽しくてシンプル、それでいて情報が満載された、作品の展示スペースです。これがあれば、顧客に作品集を送付する必要もありません。作成において最も頭を痛めたのは、Webサイトのデザインと作品の配置です。ユーモアと簡潔さを表現し、違和感のない方法で作品を見てもらうために、パラパラと本をめくるときのテンポを再現しました。ここでは、自分の速度で作品を閲覧し、どこからでも好きなコンテンツへ移動することができます。Webサイトのデザインは初めてだったため、思い通りに完成させるのに苦労しましたが、できあがるととても気に入りました。

The main objective was to make a site that was fun, simple, and informative—a place where I can show people a nice cross-section of my work, so I can avoid the hassle of sending a portfolio. The biggest challenge was deciding on a final design and narrowing the focus of the work. I really wanted to convey a sense of humor and simplicity. I wanted to present my work in a way that people are used to seeing it. When you show people your book, they generally just flip through it; that was the tempo I tried to duplicate. The site allows visitors to go at their own pace and jump around in a non-linear fashion. This is my first site design, which makes it extra hard to ever be satisfied. But I'm really happy with the end result.

● **fontgraphic.com**
● **http://www.fontgraphic.com/**
● Japan／1999

CL, SB, DF：大谷秀映 Hideaki Ohtani

concept コックピット型インターフェイス。フォントを核としたウェブデザイン制作。

Fonts are central to this Web design site; some can be accessed from the cockpit interface designs shown at the site.

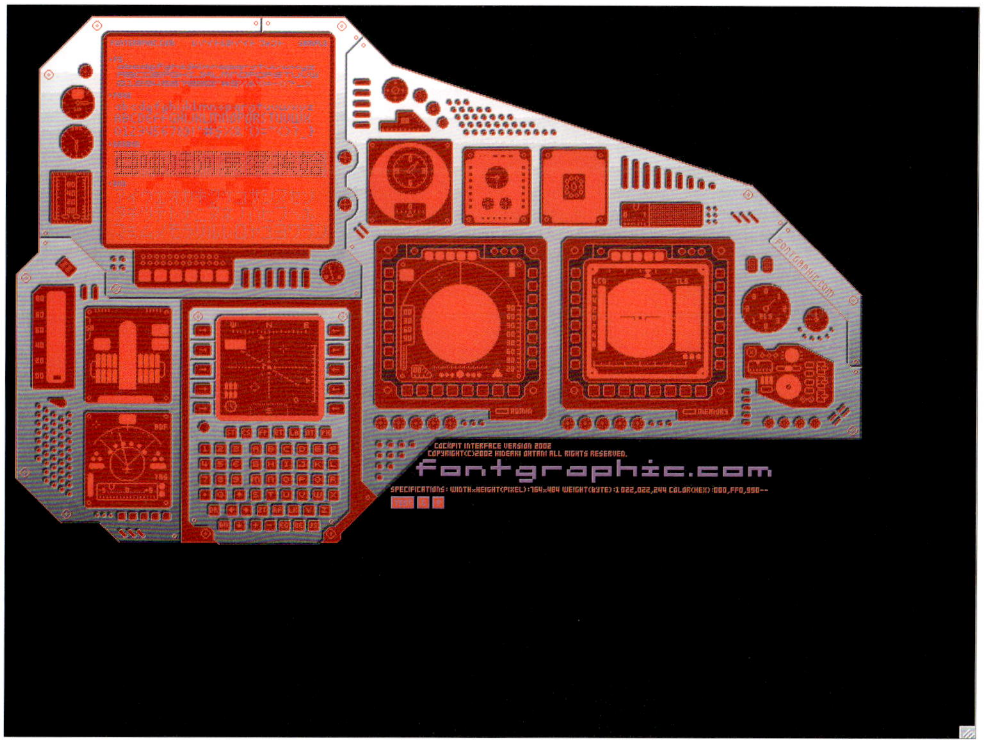

● **theBrainbox - TVAddicted**
● **http://www.thebrainbox.com/projects/tvaddicted/index.html**
● **ITALY／2002**
CL, SB：theBrainbox DF：Mauro Gatti and Lorenzo Manfredi

concept 従来のプラットフォームや戦闘ゲームとは一味違うゲームを作成しました。指を使うだけのシンプルな操作で、予想もしない展開が用意されています。

The idea was to create a funny game unlike classic platforms or fighting games. The game is simple to play—you just use one finger—but has an unexpected conclusion.

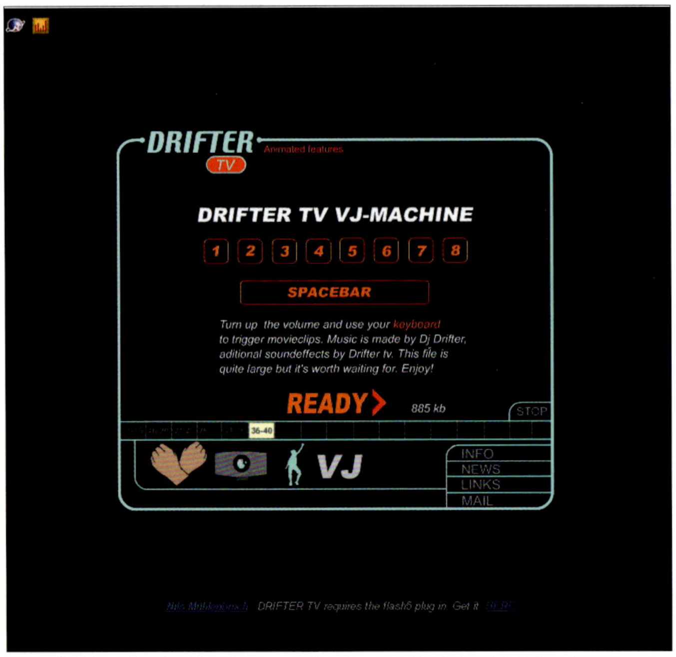

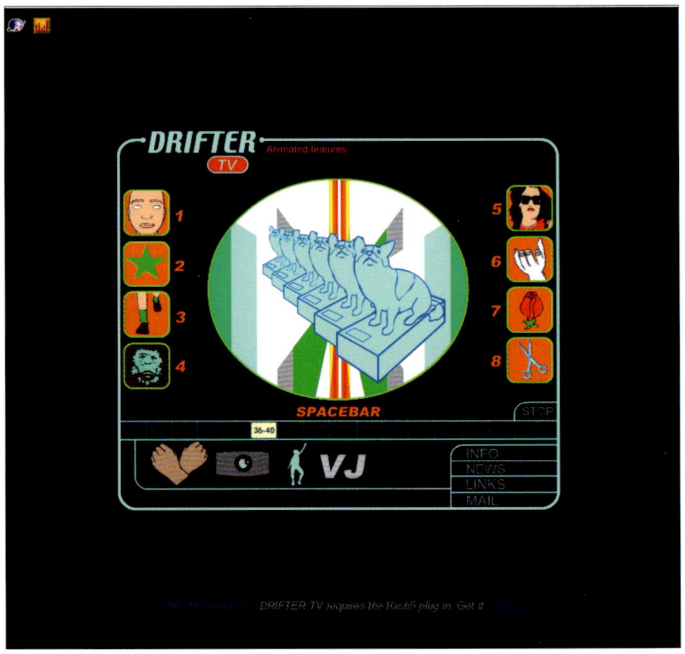

● **DRIFTER TV** ● **http://www.drifter.tv** ● **THE NETHERLANDS／2000**
CL, SB：DRIFTER TV DF：Nils Muhlenbruch

concept 短編のアニメーションを発表するWebサイト。ムービーの中にはクリックすることができるものもありますので遊んでみてください。ムービーに登場する怪しいいでたちの「ドリフター（放浪者）」は、人間を助け出すためにこの世に送り込まれてきたはずだったのですが、自分のことでもっぱら忙しい様子。作品に簡単にアクセスでき、訪問者と作品を共有できるオープンなスペースとして2000年はじめに立ち上げた個人サイトです。

Drifter.tv is a Web site that shows short animated features; some of these movies are clickable, some of them are not. A lot of my movies depict scout-like figures called drifters. They were sent down to earth to help us out, but I guess they are quite occupied with themselves as well. It is a personal site that I started up early in 2000. I wanted to create an open space for my artwork that was easily accessible and in a medium that would give the audience the opportunity to interact with the artwork.

● **DROPPOD** ● **http://www.droppod.com**
● USA／2002
CL, SB：Droppod DF：Mat A. Mejia

concept ロサンジェルスを拠点に活動するデザイナー、Mat A.Mejiaの個人サイト。最新かつ最小限のスタイルとレイアウトをコンセプトとしています。「work」のナビゲーションでは、データのロードプロセスを思わせるデザインと無機的なサウンドが使用されています。現在、カリフォルニア州ビバリーヒルズのDNAスタジオでシニア・アート・ディレクターを務めるマットは1996年以降、インタラクティブデザイナーとして、コカコーラ、ホンダ、ユニバーサル、ソニー、NASA、ABC、トヨタ、Anheiser-Busch、Foxなど多数の企業に向けて作品を提供しています。

Droppod is the personal site of Los Angeles-based designer Mat A. Mejia. The concept behind Droppod was to execute a contemporary but minimalist style and layout. The "work" content navigation was built to stylistically mimic the process of data loading, and is complemented with static audio.Mat is currently the Senior Art Director at DNA Studio in Beverly Hills, California. He has worked as an interactive designer since 1996, helping companies such as Coca-Cola, Honda, Universal, Sony, NASA, ABC, Toyota, Anheuser-Busch, and Fox achieve their business and communication objectives.

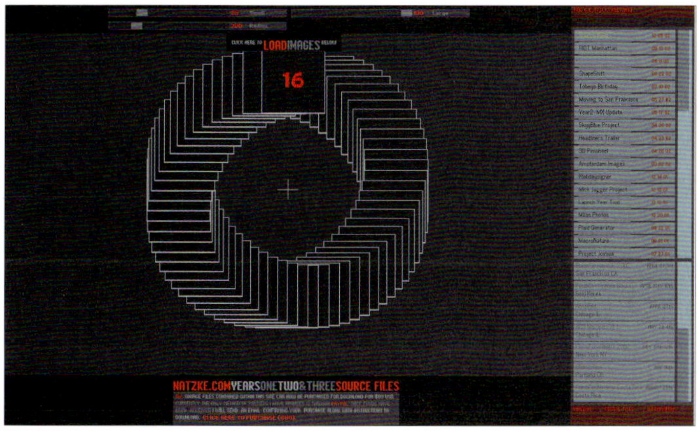

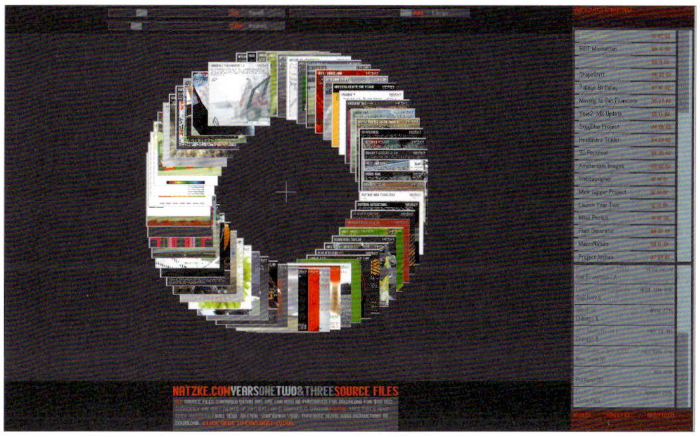

● **Natzke.com** ● **http://www.natzke.com/** ● USA／2002
CL, SB, DF：Erik Natzke

concept 私のサイト用メニュー・システムは、私がインターフェースアイデア(一般にマクロメディア・フラッシュの適用の使用を備えた)を発展させて行った調査を表わす一連の研究カタログを作る必要から構築されました。主要なポイントは、ダウンロードすることが迅速で、しかも使用することが容易なものを作成することであり、それに基づいて構築しました。メニューはフラッシュMXを使用することにより、外部のxmlファイルから離れて、ダイナミックに構築されています。また、theentireファイル(あるいは初期のダウンロード)は10k未満です。

The menu system for my site was built out of the necessity to catalogue a series of explorative studies of ideas for interfaces—most use Macromedia's Flash. I focused primarily on creating a menu that is easy to use, downloads quickly, and can build itself dynamically based on content. I used Flash MX to construct the menu based on an external http://www.xml file. The initial download is less than 10k.

● **Design & Fine Art** ● **http://www.redaart.com/** ● **USA／2002**
CL, DF：Salvatore Reda SB：salvatore reda design

concept ナビゲーション
「ある場所から別の場所へのコースを図表化し、追跡する技術、および移動中の船舶、航空機、その他の乗り物の位置を判別する技術・学問」
ナビゲーションは、与えられた素材を読み取り、理解する上で重要な要素です。私の作品はモビリティがコンセプトです。訪問者は常にコンテンツと向き合い、個々の判断で情報の位置づけを決定するものです。情報のパラメータをいかに面白く挑発的に伝えるか。その手段をつくりあげるのが私の仕事です。

Navigation is defined as "the science of plotting and following a course from one place to another and of determining the position of a moving ship, aircraft, or other vehicle."
Navigation is an important aspect in the ability to read and understand source material. The viewer must be able to navigate through the material easily and quickly. Mobility is the concept that inspires my work. Viewers position themselves in front of information I provide and reach their own conclusions. My job is to create interesting and challenging experimental forms for information parameters.

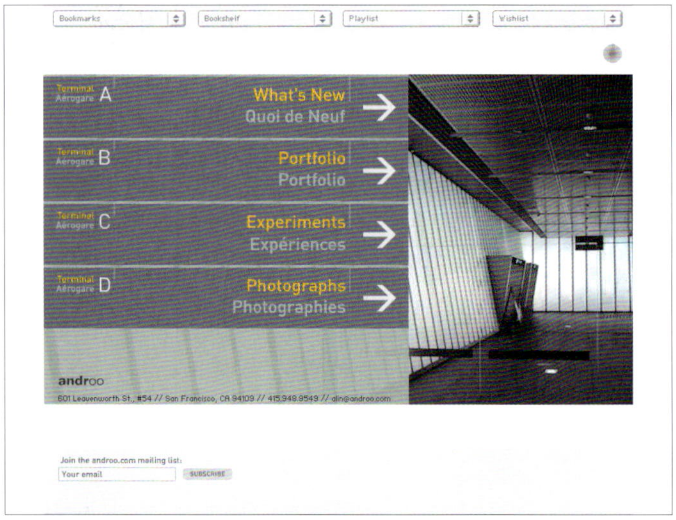

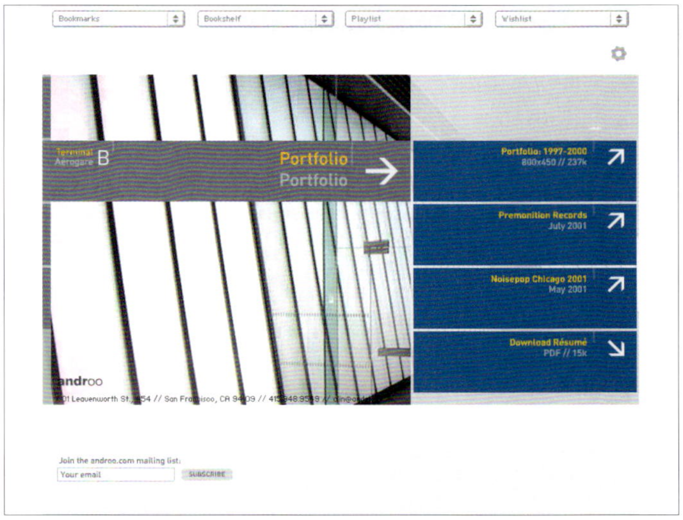

● **Androo.com** ● **http://www.androo.com** ● **USA／2001**
CL, SB, DF：Andrew Lin

concept Anderew Linのセルフプロモーション用Webサイト。未発表を含む作品が掲載されています。空港などの案内標識をモチーフにデザインされています。

Androo.com is a self-promotional Web site that presents Andrew Lin's personal and professional work. The site is seamlessly navigated with a menu that imitates directional signage found in airports and other public spaces.

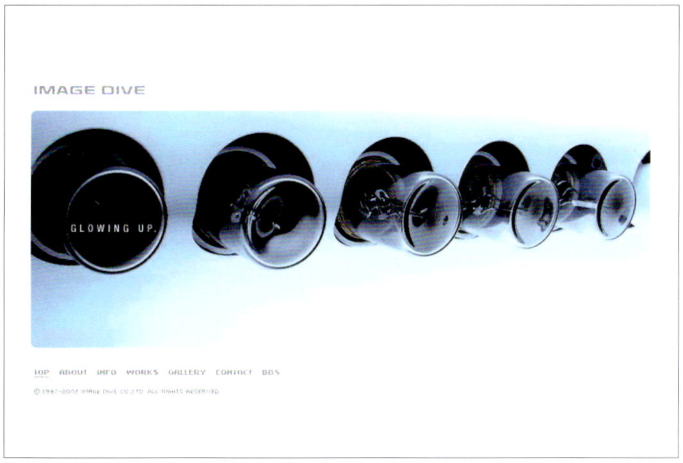

● **IMAGE DIVE** ● **http://www.imagedive.com** ● **JAPAN／2001**
CL, SB, DF：(有) イメージダイブ IMAGE DIVE CO., LTD.

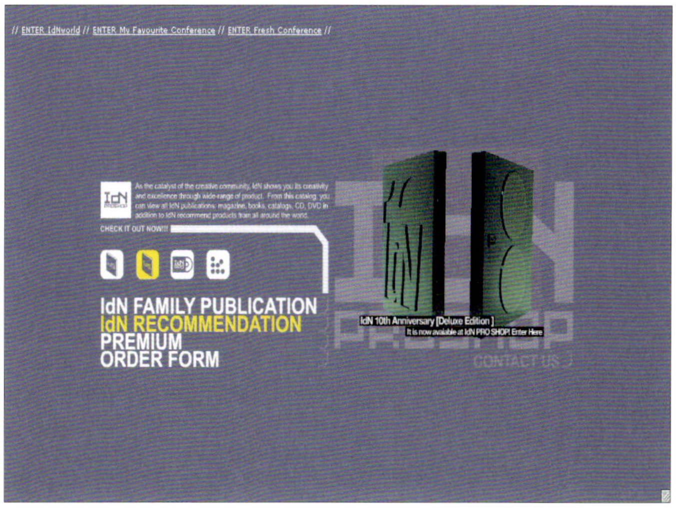

● **IdNWorld.com** ● **http://www.idnworld.com** ● **CHINA／1994**
CL, SB：IdN DF：IdN design team.

concept
オンライン書店。デザイナーの製品も数多く紹介されています。

The IdN online bookstore also carries lots of interesting design objects.

● **UNseeN Magazine** ● **http://www.unseenmagazine.com** ● **AUSTRALIA／2001-2002**
CL, SB：UNseeN DF：Garry Trinh

concept
奇抜なアイデアを具現化し、発表する場所です。真夜中にふとひらめいたアイデアを発表する私の遊び場です。

UNseeN is a place for me and others to create and publish quirky ideas. Themes and ideas come to me in the middle of the night and I use UNseeN to bring them to light. It is my playground.

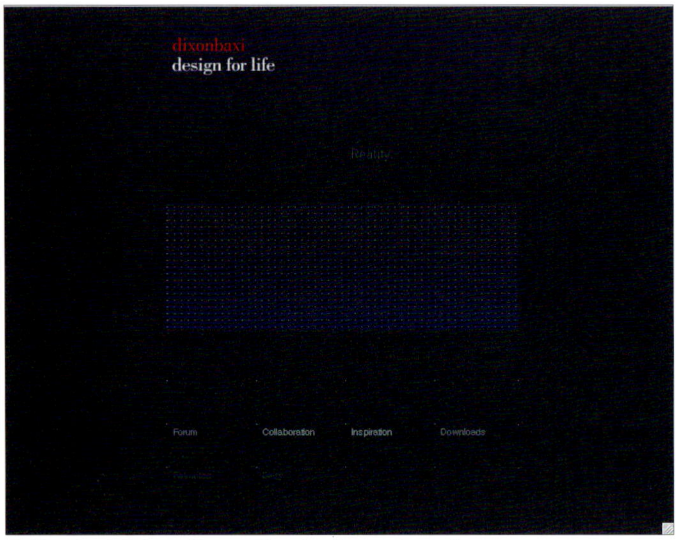

● dixonbaxi ● http://www.dixonbaxi.com ● UK／2002
CL, SB, DF：dixonbaxi

concept 2002年1月に設立されたロンドンのデザイン事務所のために、数日で作成したWebサイトです。作者がこれまでに手がけてきた「見せる」Webサイトとは対極的に、意識的に何も見せないというコンセプトを打ち出しています。デザイン事務所らしからぬと思われるかもしれませんが、その理由は訪問者の注意をコンテンツに引き寄せることにあります。部分的にはFlashを使用していますが、全体としてはテキストをベースとしており、デザインを極力抑えました。トップページにはわずかな文字以外に何も表示されないため、訪問者は必然的にサイトに目を向けるようになります。作者が「会話」と呼ぶこのページは、不定期に更新しています。実際にWebサイトの中に入るには、ボタンを探し出さなくてはなりません。いったんボタンを見つければそこから先は簡単です。クリックすれば、テキストのコンテンツが表示されます。このWebサイトは、デザイン事務所の成長記録でもあります。サイトを日記帳に見立て、進行中のプロジェクトなどについて公開します。訪問者が読み応えを感じるコンテンツを目指しています。画像もなく事実を羅列しているだけですが、読者からの反応は好評です。また、一流のプロとデザインについて議論する場でもあります。当初は訪問者向けのコンテンツを発表する場でしたが、その後コラボレーションに発展することもありました。
現在、このサイトは再構築中です。美的感覚とトーンを維持したまま、コンテンツを整理し、新機能も追加する予定です。今度は画像が挿入されるようになるかもしれません。

We created this site in a matter of days so that our London-based design practice, launched in January 2002, would have a Web presence. Although we have plenty of work to show we made the conscious decision not to show any. A strange decision for a design practice, you might think. It is the written content, however, that we want to focus on and direct the user's attention to. So the site, although created in Flash, is entirely text based. The design is extremely restrained; when visitors arrive at the site they see only a few words at first. We call this the "conversation," because that's what it is. The conversation is updated at irregular intervals. To display the content, the user first has to discover the buttons. After that it's simple—just click the buttons to bring up text. The site was constructed with the aim of charting our progress as a growing design practice. We use it like a diary to candidly record what we are up to. Hopefully, it will prove interesting reading. Judging by responses, the site does seem to stimulate discussion—if only about the fact that there are no images.The site also serves as a forum where we invite respected industry professionals to discuss specific design topics. This generates a fascinating pool of content for visitors to explore and has led to some interesting collaborations.We are currently restructuring the site. The basic aesthetic and tone will be preserved, but content will be streamlined and some extra features added. We may even put up a few images...

● mediumrare ● http://www.mediumrare.net ● UK／2001
CL, SB：Mediumrar DF：Christian Nouyou

concept コンセプトは「シンプル・イズ・ベスト」です。
Simple is best!

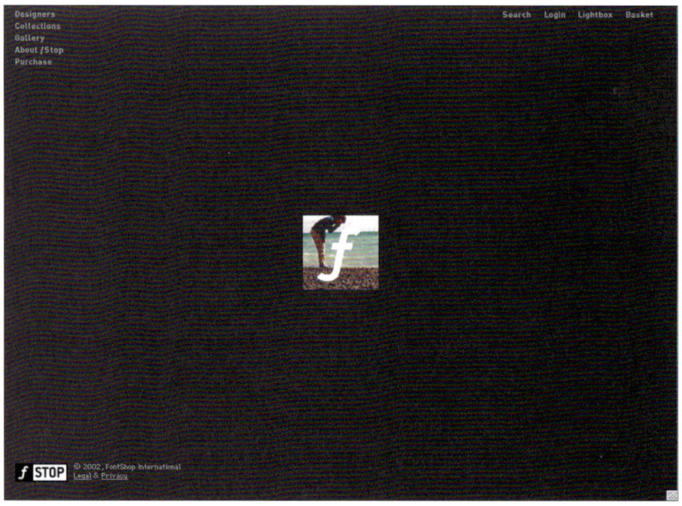

● fSTOP ● http://www.fStopimages.com/ ● USA／2002
CL, SB：FSI USA

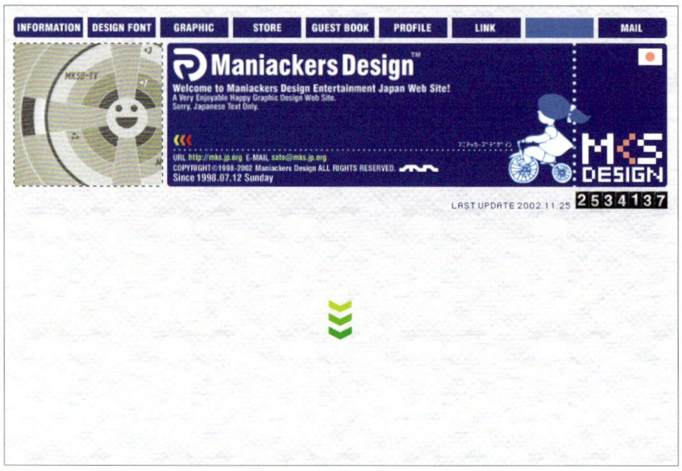

● Maniackers Design ● http://mks.jp.org ● JAPAN／2001
CL, SB, DF：Maniackers Design

● TURUX SEMI-AUTOMATIC. ● http://www.turux.org ● AUSTRIA／1997
CL, SB：turux.org DF：dextro

concept インタラクティブまたはランダムなイラストやサウンドを生成するツールを提供しています。

The site provides interactive and/or random tools for picture- and sound generation.

● This is a Magazine. ● http://www.thisisamagazine.com ● ITALY／2002
CL, SB：karen+andy photo+graphics DF：andy simionato

concept 最新ビジュアル言語のWebサイト。管理と編集機能を備えた、未知の領域へのビジュアルコミュニケーションを可能にするギャラリースペースです。

The site is a critical container for contemporary visual languages—a gallery space that we curate and participate in, where visual communication can be taken to unexplored areas.

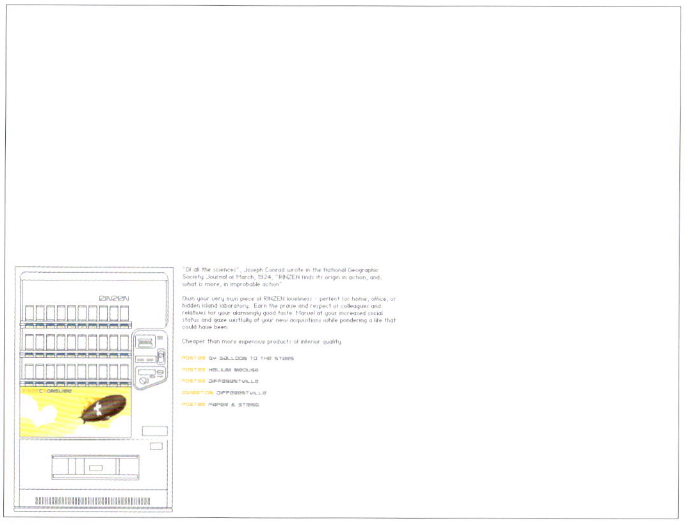

● RINZEN ● http://www.rinzen.com/themachine.html
● AUSTRALIA／2002
CL, SB, DF：RINZEN

● Nobody Here ● http://www.nobodyhere.com/
● THE NETHERLANDS／2002
CL, SB：Nobody Here

● EDITOR ● http://www.labmagazine.co.uk ● UK／2000
CL, DF：ARTMIKS SB：PAVLOVA & LAB MAGAZINE LTD.

concept Fischer-Priceの子供用ゲームをモチーフにしたWebサイト。2002年12月末オープン予定です。

The Web site is inspired by Fischer-Price children's games. Now there is just a splash page; the site is currently being built and will launch at the end of December 2002.

● The site of yooco ● http://www.yooco.com ● JAPAN／2001
CL, SB：谷本ヨーコ Yooco Tanimoto DF：浜田武志 Takeshi Hamada

concept 自分の作品のポートフォリオ的な目的で制作しています。ですが、ただ仕事の作品を並べるのでは見る側も飽きてしまうのでアニメーションを使った女の子のキャラクターのアイコンを全体に通して使ったりと楽しい要素を入れています。

I created the Web site to be a portfolio presentation of my artwork. But I imagine viewers might get bored if all they saw was just pages and pages of my work, so I put an animated female character icon on all the pages to inject some fun into the site.

● ya Dig it ? ● http://www.yadigit.org/ ● 2002
CL, SB：Ya Dig It ?

● **YFACTOR** ● **http://www.yfactor.us** ● **ISRAEL／2002**
CL, DF： yonatan factor SB： yfactor

concept ビジネスに関係なく、個人的な作品を展示するWebサイトです。さまざまな方法（単純で直接的な伝達方法、遊び心いっぱいの色やフォーム、特定テーマの明確な解釈など）で試験的にメッセージを発信しています。ここでは、できるだけ既製の素材は使用せずにオリジナルの素材を活用しています。サイトに掲載されているイラストやテキストなどはすべて同一人物によるものです。作品は主に音楽、文化、社会をモチーフとしており、すべて私的に自由に創作したものです。仕事から離れた創作であるため、直感でひらめいたテーマから熟慮して決定したテーマを幅広く取り上げ、固定観念に縛られることなく自由につくりあげることができました。アートワークは本来、印刷物です。Webサイトは印刷の位置づけを確認するために作成しました。したがって、シンプルな構成と余裕のあるスペースが必要です。何かが飛び出したり、移動したり、揺れ動く必要はありません。あくまでイラストが中心です。Webサイトにイラストを展示しているのではなく、イラスト自体がサイトなのです。技術的にWebを活用しただけなので、ネット上で閲覧している感をできるだけ払拭するよう工夫しました。主役である作品の邪魔をしないよう、構成をクリアにし、飾り気のない実用的なナビゲーションバーを採り入れています。

This Web site is a collection of several independent personal works, without any commercial intent. They form a laboratory of experiments in transmitting diverse messages in a variety of ways—simple and direct communication, playful use of color and form, and clear interpretation of a particular theme. My intent and desire is to make maximum use of personal and original materials and minimize ready-made materials. The illustrations and the text, along with the rest of the elements in this site, are the products of the same signature. Music, culture, and society are central to a large portion of my work. The illustrations featured in the site are independent; they were not commissioned by clients. The lack of a commercial goal gives me the liberty to charge at different themes and subjects impulsively or deliberately, and to work freely with no interruption. The artwork is essentially print-based. I wanted the site to serve the print orientation and not the other way around, so it is simple and sparse. Nothing pops up, moves, or wiggles. No Flash manipulations are employed. The site is a vessel for transmitting messages. I wanted to focus on the illustrations—the illustrations are the site—and minimize the Internet feel. They are exhibited there for purely technical reasons. Clear, basic site division and a practical, unpretentious navigation bar give maximum respect to the work itself.

● **MONSTERISM** ● **http://www.monsterism.net/** ● **UK／2002**
CL, SB： Pete Fowler DF: BROAD SNOUT

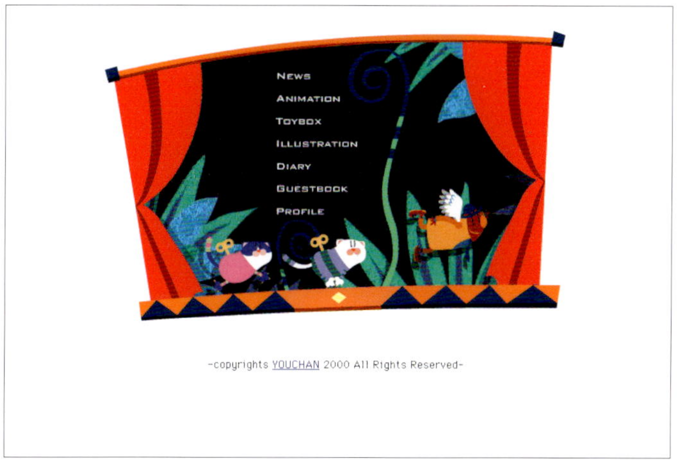

● **YOUCHAN** ● **http:// www.youchan.com/** ● **JAPAN／1998**
CL, SB, DF： YOUCHAN (TOGORU Company)

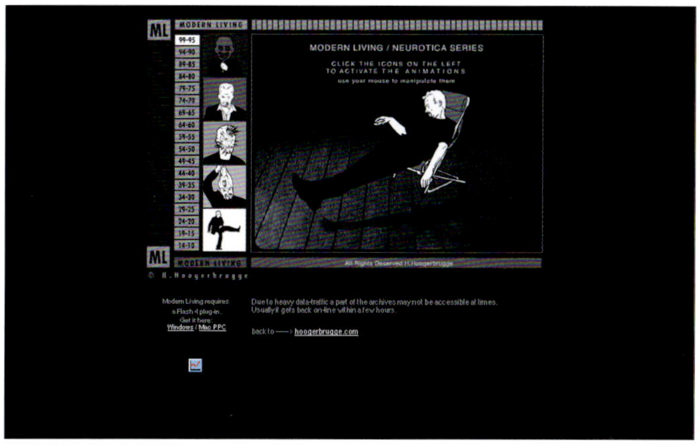

● **MODERN LIVING**
● **http://www.hoogerbrugge.com/ml.html** ● **THE NETHERLANDS／1998-2001**
CL, DF： Han Hoogerbrugge SB： Hoogerbrugge.com

● **FLAT** ● **http://www.flat.com** ● **USA／2002**
CL, SB, DF： Flat

concept ニューヨークに拠点を置くFLAT（1997年設立）のWebサイト。複雑かつインタラクティブなデザインシステムに注目するFLATは、インタラクティブデザイン、グラフィックデザインおよびコンテンツ開発を基幹事業とし、これまでに数多くのプロジェクトに携わってきました。その分野はオンラインマーケット、Webサイト公開、インタラクティブゲーム、コーポレートアイデンティティパッケージ、パンフレット・カタログ、組立式家具システムのほか、おしゃべり人形まで多岐にわたります。

This is the Web site for FLAT, a design firm based in New York City. Founded in 1997, FLAT focuses on complex interactive and design systems. FLAT's core competencies are interactive design, graphic design, and content development. We've worked on a diverse set of projects so far—online marketplaces, Web site publishing, interactive games, corporate identity packages, print collateral, a modular furniture system, and one talking doll.

● **GAUCHO** ● **http://www.gaucho.com/21/** ● **JAPAN／2002**
CL, SB, DF： WEB DESIGN STUDIO GAUCHO

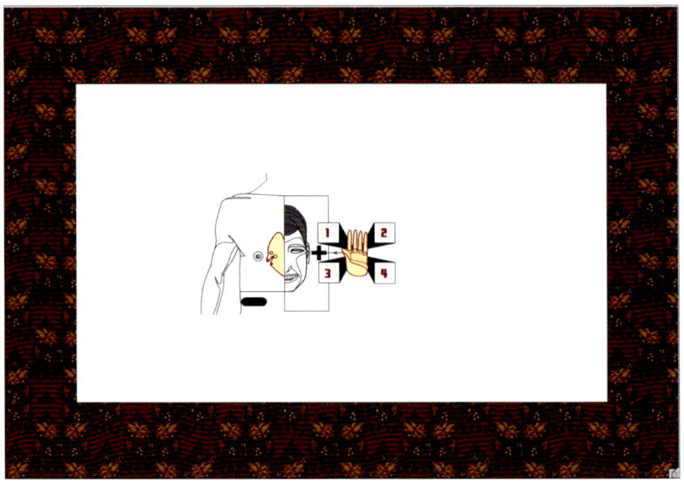

● **CISMA** ● **http://www.cisma.com.br** ● **BRAZIL／2001**
CL, SB, DF： CISMA

● **James Jean Illustration** ● **http://www.jamesjean.com** ● **USA／2001**
CL, SB, DF：James Jean

concept 「単純明快で便利」がコンセプトのイラスト展示サイト。気品や芸術性を損なわずに、必要な要素や伝えたい情報を1つのウィンドウに収めるよう工夫しました。絵画や彫刻の制作課程と同じように、それぞれの要素と枠（ここではウィンドウ）がほどよく調和するよう、レイアウトを工夫しました。

I stressed simplicity, clarity, and convenience in the creation of my illustration portfolio site. I tried to keep the essential elements and information in one window whenever possible, and also produce an elegant and beautiful site. Just as in drawing or sculpture, I experimented with different proportions until I found a comfortable relationship between each element and the limits of its bordersŃin this case, the window.

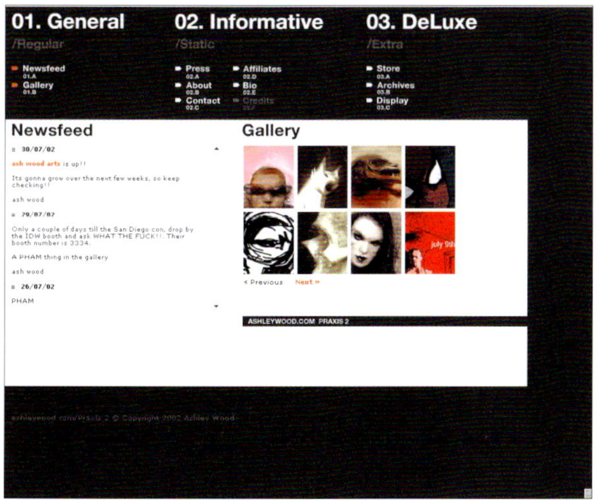

● **ASHLEYWOOD.COM** ● **http://www.ashleywood.com/praxis2/**
● **USA／2002**
CL, SB：Ashley Wood

● **SPEKTRUM STUDIOS** ● **http://www.spektrumstudios.dk/**
● **DENMARK／2002**
CL, SB：Spektrum Studios

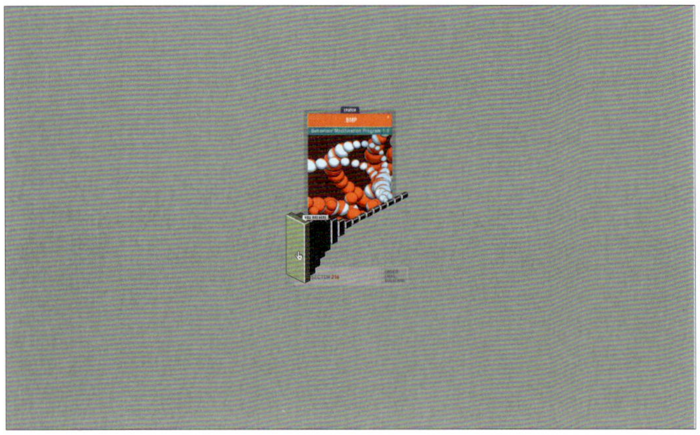

SECTOR216 **http://www.sector216.com** **THE NETHERLANDS／1999**

CL：www.superfamous.com / www.fluid.nl SB：Sector216 DF：Remon Tijssen / Folkert Gorter

concept Remon Tijssen（Fluid.nl）とFolkert Gorter（Superfamous.com、Newstoday.com）のコラボレーション作品を公開するWebサイトです。私たちはこれまでに共同で数多くの作品を完成させてきましたが、試作品や個々の作品を展示する場を持っていませんでした。それを実現したのがSector216です。
ここは創造力を発表するスペースです。新しいアイデア、テクノロジー、デザイン、インターフェイスはここで生まれます。つまり、個々の創作活動に不可欠な存在なのです。その効果として、コラボレーションの結果が単なる「1+1=2」の図式に終わることはありません。私たちはいつも、シンプルなアイデアが浮かぶと何十回でもバラバラに分解してみたり、大きく拡大したりといったやりとりを繰り返して作品を完成させます。どんなプロジェクトでも、自然のままの動き、遊び心、対話性を採り入れることにしています。現在普及しているデジタル式の情報伝達方法には改善点が多く、一味違ったものをつくるには絶好のチャンスです。Sector216にはゴールはありません。コンテンツを定期的に追加し、時代に応じた方向性を持たせていきたいと思っています。今後の展開にご注目ください。

Sector216 was initially created to accommodate cross-continent collaboration between Remon Tijssen (fluid.nl) and Folkert Gorter (superfamous.com, newstoday.com). We have been working together quite a while, but never had a stage or an audience for our more experimental and personal work. Sector216 fills this void. The idea is to use the space as a creative outlet—a breeding ground for new ideas, technology, design and interface behavior—that is essential to our personal development. We've found that our collaboration seems to produce 1+1=3 results. Our projects usually start out as simple ideas, and we may pass them back and forth as few as 4 or as many as 40 times, breaking and stretching them in the process, until we feel we have a solid piece ready for public viewing. Natural behavior is an element we inject into each of our projects, along with a sense of playfulness and awkwardly familiar interactivity. There is so much to explore and improve upon, and a real chance to make a difference in the way people experience digital information and entertainment. Sector216 does not really have a set goal; we will keep adding to it and let it have it's own life—see where it goes and what will appear along the way.

Re:move **http://www.re-move.org** **1999-2002**

CL, SB, DF：lia

concept このWebサイトにはテキストを使っていません。ページから発せられるメッセージに耳を傾けてください。

I 'm sorry not to give a text description of the page, but it speaks for itself.

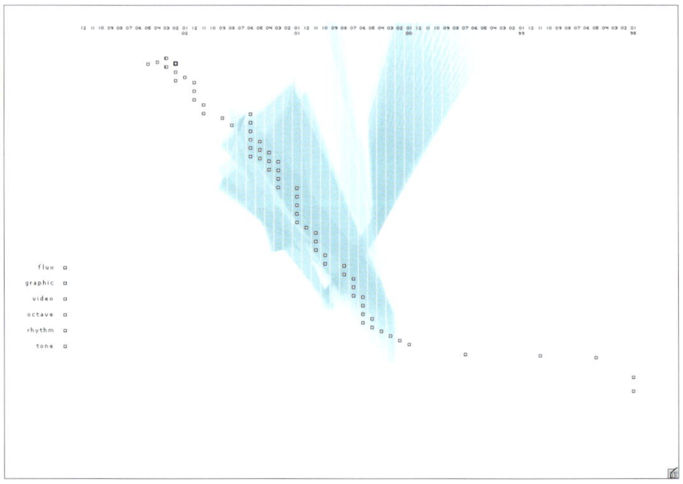

meta **http://meta.am/** **2000**

CL, SB, DF：meta

● **LOOPLAND**
● **http://www.loop-land.co.uk** ● **UK／2002**
CL, SB, DF： Allan Sanders

concept ロンドンを拠点に活動するイラストレーター、Allan SanderのWebサイト。本来、サンダーのオンライン作品集として立ち上げられましたが、現在ではアニメーション機能が試験的に追加されているほか、「Looplab」というオンラインアートプロジェクトも公開されています。

Loopland is the home of Allan Sanders, a freelance illustrator based in London. Loopland is primarily an online portfolio that showcases his commercial illustration work. The site also features some experimental animation, and online art projects in the Looplab section.

● **Joshua Davis / PrayStation** ● **http://www.joshuadavis.com/**
● **USA／2002**
CL, SB, DF： Joshua Davis / PrayStation.com

concept たとえ絵筆でキャンバスに画を描いていなくても私は画家なのだと Jackson Pollockはエッセイの中で語っています。同じように、たとえプログラム（マシン）の出力を制御できなくても、私は正統派のアーティストなのだと信じたいと思います。私の仕事は、ペイント、ブラシ、キャンバス、キーボードなどのツールを使って罫線や枠を入力することで、それらを出力するのはマシンです。マシンとの共同作業は驚嘆と発見の連続です。私の意図どおりに出力されることは決してないのですから。

Jackson Pollock said in some essays that he considered himself a painter, though at times the brush never hit the canvas. Similarly, I would consider myself a traditional artist, though I have no control over what my programs (machines) output. I input the paints, brushes, canvas, strokes, rules, and boundaries, but it is the machine that outputs the compositions. So I, the artist, am in a constant state of surprise and discovery, because the machine may construct forms I never imagined.

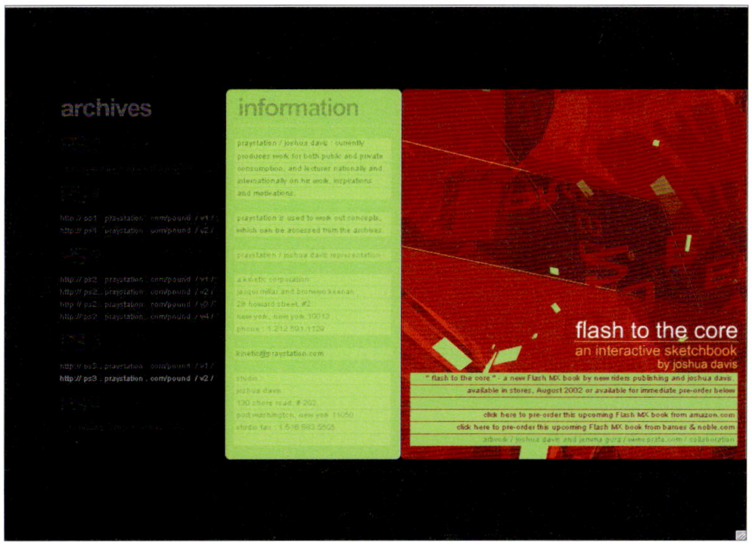

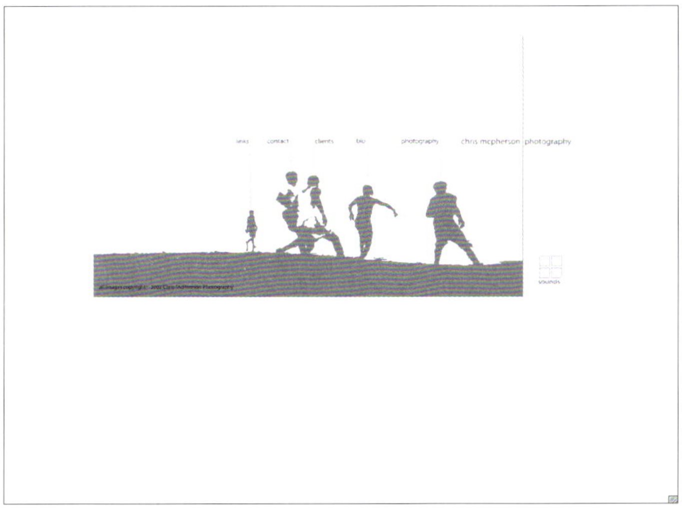

● **Chris McPherson Photography**
● **http://www.chrismcpherson.com/main.html** ● **2002**
CL, SB： Chris McPherson Photography

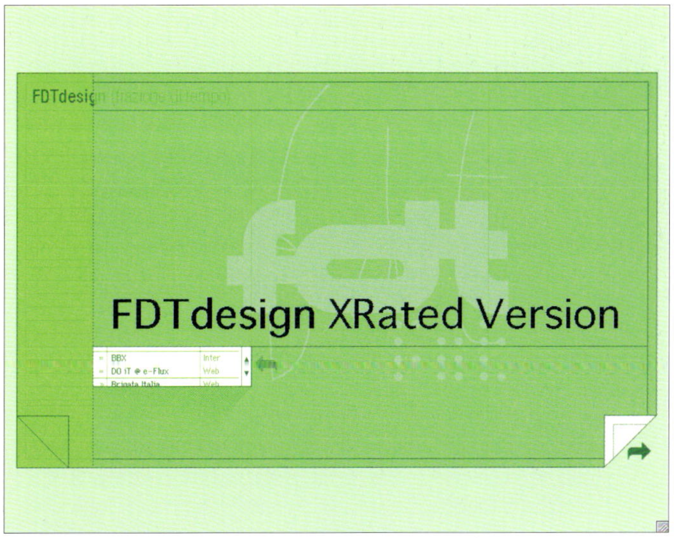

● **FDTdesign** ● **http://www.fdtdesign.com** ● **USA／2002**
CL, SB： FDTdesign DF： Marco Moretti / Ty Leung

● Ocular ● http://www.ocular.com ● USA／2002
CL, SB：Ocular DF：Geoff Lillemon

concept Ocularの作品には、無垢なる革命という背景で定義された芸術を生命に反映するというコンセプトが刻まれています。芸術へ哲学的にアプローチすると、人間は創造者ではなく、しょせん革命者にすぎず、最も崇高な革命は先入観を完全に排除した無垢であると理論づけられます。人間がいくら自分自身を創造者である、または何らかを創造していると位置づけてみても、実際には、いつのまにか何かの意図を持って与えられた道具を使って物事を「改良」しているだけなのです。人間自体は自然の万物の一部に過ぎません。精神の鍛錬を重ねたとしても、すばらしい成果が得られるのは稀で、創造に関連することに携わっているだけなのです。
例を挙げてみましょう。ダーウィンの進化論によると、地球上に生命を発生させた原始スープから進化した人間は、自然界からコンピュータのようなより高度な道具を採り入れ、その道具を利用して最も崇高な祈りの形である芸術を形成し、純真に生命を賞賛しています。アリは小枝を使って小川をわたります。無数の小枝を重ねれば、やがて橋になります。アリが優れた頭脳を駆使して成果を得ていることは評価できますが、このプロセスを創造とみなせるでしょうか。アリは無垢に行動しているだけです。難局に面していることも、難局を打破する道具も意識せず、ただ単に生きる環境を改善するために一致した行動を見せているだけと想定できないでしょうか。
Ocularの作品は、芸術の反映である生命に調和して無垢なる革命を起こしています。

Life as the reflection of art defined within the context of innocent innovation is the concept that permeates work at Ocular. Our philosophical approach to art theorizes that man is not a creator but only an innovator, and innovation in its highest form is innocent and without true preconception. This means that even when we consider ourselves to be creators or to be creating, we are actually just making things "better" with tools that have already been purposefully and naturally given to us. Man is a natural resource himself. Even in the exercise of his mental abilities, he is at best a distant—although sometimes brilliant—cousin to creation.
To illustrate the point, consider the following analogies. Man evolved in the Darwinian fashion from primordial ooze and moved up to the stage of being able to innovate computers and other refined tools from nature, then took another step up to innovate art—the highest form of prayer—from these refinements, in childlike celebration of life itself. An ant uses a twig to cross a stream; it does this so many times that the twigs become a bridge. At no point in the process is the ant considered a creator, even though it assessed the situation and came up with a conclusion that snowballed into brilliance. One assumes that the ant acted innocently and neither created itself, its dilemma, or the tools to solve the dilemma, but simply acted in unison with life to better its situation.
Ocular is innocent innovation in unison with life as a reflection of art.

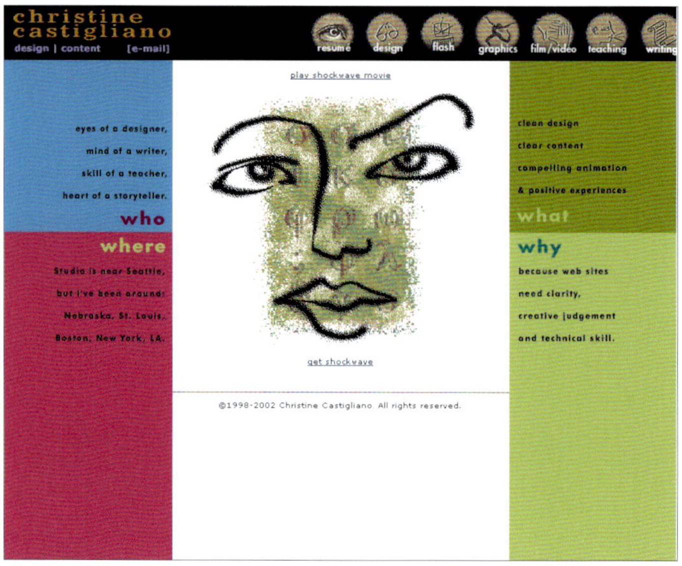

● metaToggle ● http://www.metatoggle.com/home.html ● USA／1998
CL, SB：metatoggle DF：Christine Castigliano

concept 私は元ジャーナリストとして、Webサイトに必要なのは真実を迅速に伝達することだと考えています。作品集では、依頼者（Who）、依頼目的（Why）、作品の説明（What）、制作場所（where）の枠をそれぞれ作品の周りに配置し、詳細を説明しています。私の作品の概念だけでなく、作文技術やデザインスキルも発表できる場です。

As a former journalist, I understand the need for a Web site to quickly communicate the facts about who, what, when, and where. I structured my portfolio with four squares, one for each of the four questions. This format allows me to demonstrate my conceptual and writing abilities along with my design skill.

● James Leland Photography ● http://www.lelandphoto.com
● USA／2002
CL, SB：James Leland Photography DF：Gabe Rubin at www.wrecked.nu

concept 写真家ジェームズ・レランド（James Leland）の作品を展示するWebサイト。ジェームズは10年以上もセレブリティ、ファッションモデル、ミュージシャンを撮影し、Raygun、Bikini、Huh、『Rolling Stones』誌のほか、コールハーン（靴）、レイバン（アイウェア）、リーボック（靴）、クラフト（食品）などをクライアントとしています。
www.lelandphoto.com is a website to showcase the photography of James Leland. James has been shooting celebrities, fashion and musicians for the past 10 ten years for such clients as Raygun, Bikini, Huh, and Rolling Stone magazines as well as commercial clients Cole Haan shoes, Rayban eyewear, Reebok shoes, and Kraft foods. This site was intended for potential clients to see a portion of the portfolio to get a sense of the work without having to ship a physical book of images.

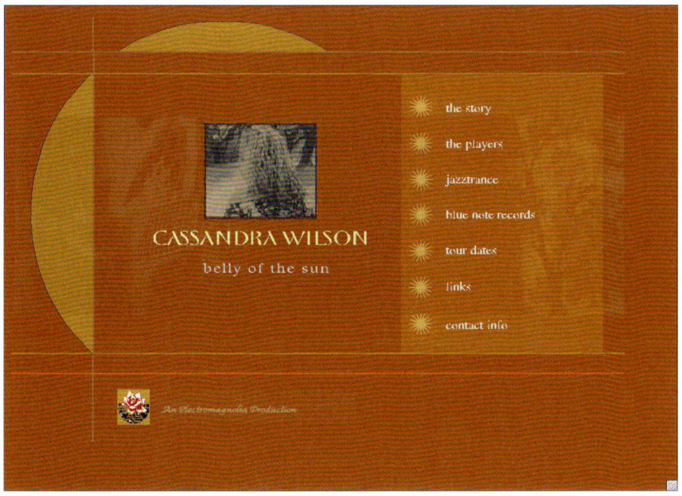

● **CASSANDRA WILSON**
● **http://www.jazztrance.com/cassandrawilson.html** ● **USA／2002**
CL. SB：An Electromagnolia Production

● **Dianne Reeves** ● **http://www.diannereeves.com/** ● **USA／2002**
CL. SB：Dianne Reeves

● **Karl Denson's Tiny Universe**
● **http://www.karldenson.com/** ● **USA／2002**
CL. SB：Karl Denson's Tiny Universe

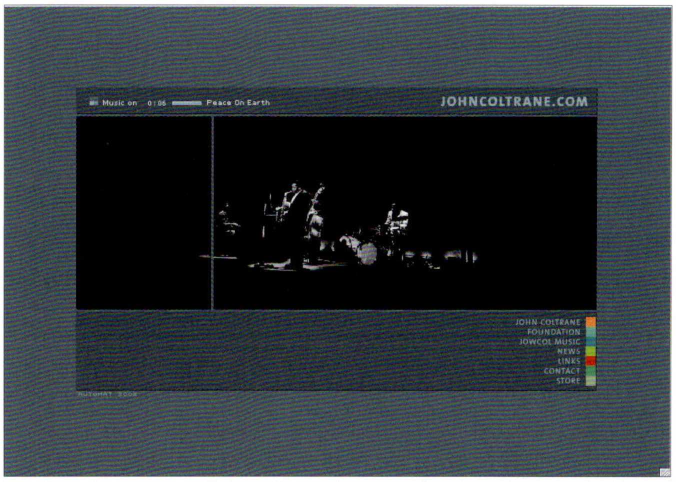

● **Johncoltrane.com** ● **http://www.johncoltrane.com** ● **AUSTRIA／2001**
CL：The John Coltrane Foundation, Inc. SB：AUTOMAT DF：Llewellen Heili, Jürgen Bauer

concept

ジョン・コルトレーンの公式サイト。ジョン・コルトレーン基金（John Coltrane Foundation）は音楽を愛する人々や彼のファンだけでなく、傑出した才能を持つミュージシャンを支援するために設立されました。このWebサイトは、天賦の才能を持つジャズサキソフォン奏者の音楽を後世に遺すために、彼の生涯が紹介されています。
サイトの中心となっているのは、彼のコレクションを紹介するライブラリです。Flashが使用されているため、ほかのメディアプレーヤーを起動する必要がありません。サウンドだけにとどまらず、Flashの効果を利用してグラフィックにも美しい動きがつけられています。ジョン・コルトレーンの功績にスポットをあてるため、あらゆるデザイン的な要素を最小限に抑えられ、一貫して調和のとれたインターフェイスが完成しました。その結果、訪問者の関心がジョン・コルトレーンとジョン・コルトレーン基金に注がれるよう工夫されています。
コンテンツは、洗練されたやわらかい色調でまとめられています。構成も概念も音楽が中心であるため、訪問者は目的のコンテンツを簡単に見つけることができます。ナビゲーションはシンプルでわかりやすく、今いる場所を色で識別することができます。音楽はすべて効率的に圧縮され、忠実に再生されます。コルトレーンのサウンドを楽しめるオーディオチャンネルは2つあり、ホームページからはバックグラウンドに流すオーディオを数曲から選択することができ、「Photos & Music」セクションでは、在りし日のコルトレーンをスライドで見ながら、傑作選を試聴することができます。

The official site of John Coltrane and the John Coltrane Foundation was created for music lovers and fans, and serves as a tribute to the outstanding musician. It presents the life of the gifted jazz saxophonist and keeps his music alive for future generations. The core of the site is its extensive music library. Flash allowed the developers to integrate a comprehensive music selection without the need for additional media players. In addition to sound, the Flash interface adds beautiful motion to the graphically spare site; all design elements are minimized so that the artist's achievements take center stage. The end result is a coherent, harmonizing interface that draws visitor attention to John Coltrane and the foundation.
The visual material is carefully organized and presented with a color palette that is appropriately subdued and sophisticated. The designers have let the musical content drive the structure and concept of the site, so visitors can easily find what they are looking for. The navigation is simple and comprehensible with a color-coded interface that functions as an information guide, reminding visitors which section they are in.
All music is efficiently compressed and reproduced with good fidelity. It is offered through dual audio channels that give site users options for experiencing the Coltrane sound. From the home page, users can choose background audio from a handful of choices. The Photos & Music section, however, offers a far greater selection of masterpieces along with a beautifully choreographed, self-running slide show.

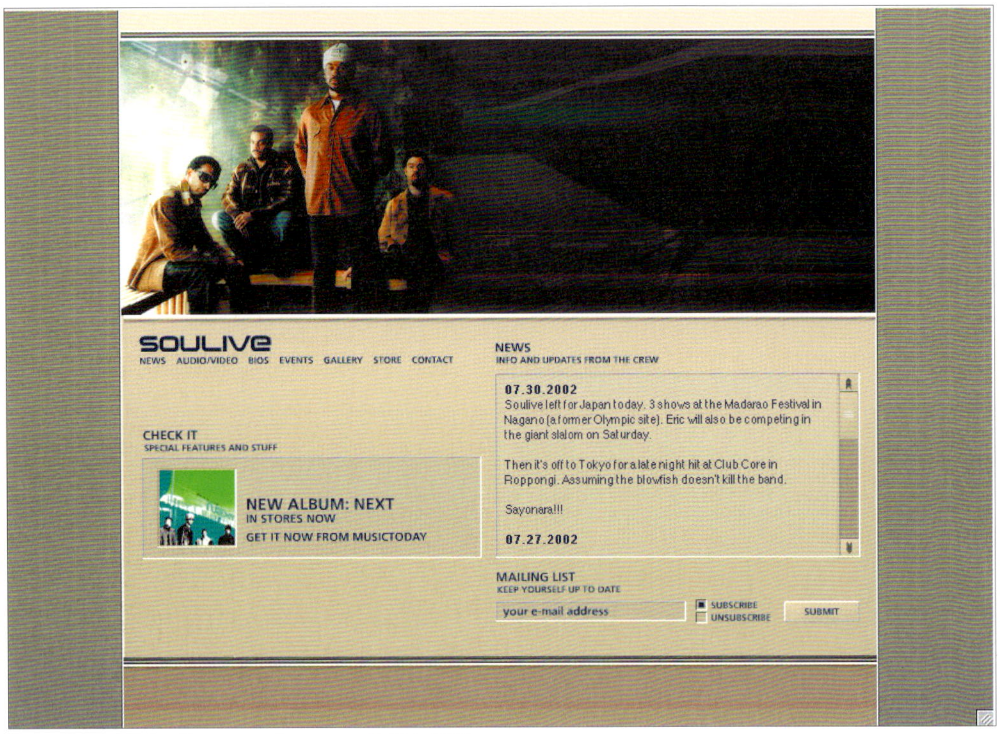

● Soulive ● http://www.soulive.com ● USA／2002
CL, SB：Soulive, Inc. DF：Andy Shaaf & Rich Bloom at Apt5a

concept ヒップホップとR&Bをジャズに融合させた音楽で世界中をツアーするグループ、SouliveのWebサイト。画像がふんだんに使われたWebサイトでは、ツアー日程などの情報を確認できるほか、楽曲を試聴することもできます。SouliveのファンはWebの知識に富み高速接続でアクセスする若年層が大半を占めることに着眼し、Flashをふんだんに駆使した構成となっています。ツアー日程、ニュース、mp3、写真は毎日更新されているため、多くのファンが定期的に訪れ、チェックしています。

Soulive, a jazz group whose music is infused with hip-hop and R&B, has recently catapulted to fame and now tours throughout the world. The Web site conveys the group's energetic image and provides fans with information on the group, tour dates, and access to some of the music. Most of Soulive's fans are young, web savvy, and have high speed connections, so there's some slick Flash stuff on the site. The site can be updated remotely, so tour dates, new information, mp3s, and pictures appear daily. This is a major factor that encourages fans to keep coming back.

● Website for jazz musician Michael Brecker ● http://www.michaelbrecker.com ● USA／2002
CL：Depth of Field Management, New York SB：Stanegate Studios DF：Geoffrey (Art Direction) / Michael Hofmann (Animation & Design)

concept 世界を代表するサキソフォン奏者マイケル・ブレッカーのWebサイト。斬新で刺激的というコンセプトのもと、Flashを駆使したアニメーション効果によりカラフルながらも繊細なナビゲーションを作成しました。また、サイト全体には動きをつけ、ブレッカーの写真を数多くちりばめました。

Michael Brecker is one of the world's leading tenor saxophonists. Stanegate Studios was commissioned to produce a modern and exciting look for Michael Brecker's new Web site. We were requested to design an animated menu/navigation banner using Flash that would be subtle yet colorful, bring movement to the site, and incorporate many different images of Mr. Brecker.

concept ファンキーですっきりとまとめられたWebサイト。わかりやすいナビゲーションでアーティストの魅力と最新リリースCDの情報を発信しています。定期的に更新されているので、アーティスト、バンド、楽曲の最新情報を確認することができます。

The site design is clean but funky. The two main objectives are: 1) to reflect the personality of the artist and introduce the current CD release, and 2) to offer fans an easy-to-navigate source of frequently updated news and information on the artist, his band, and the music.

● The Official Website of Jazz Guitarist John Scofield ● http://www.johnscofield.com ● USA／2002
CL：ScoBiz Inc. SB：Weedman Design DF：Lisa Weedman

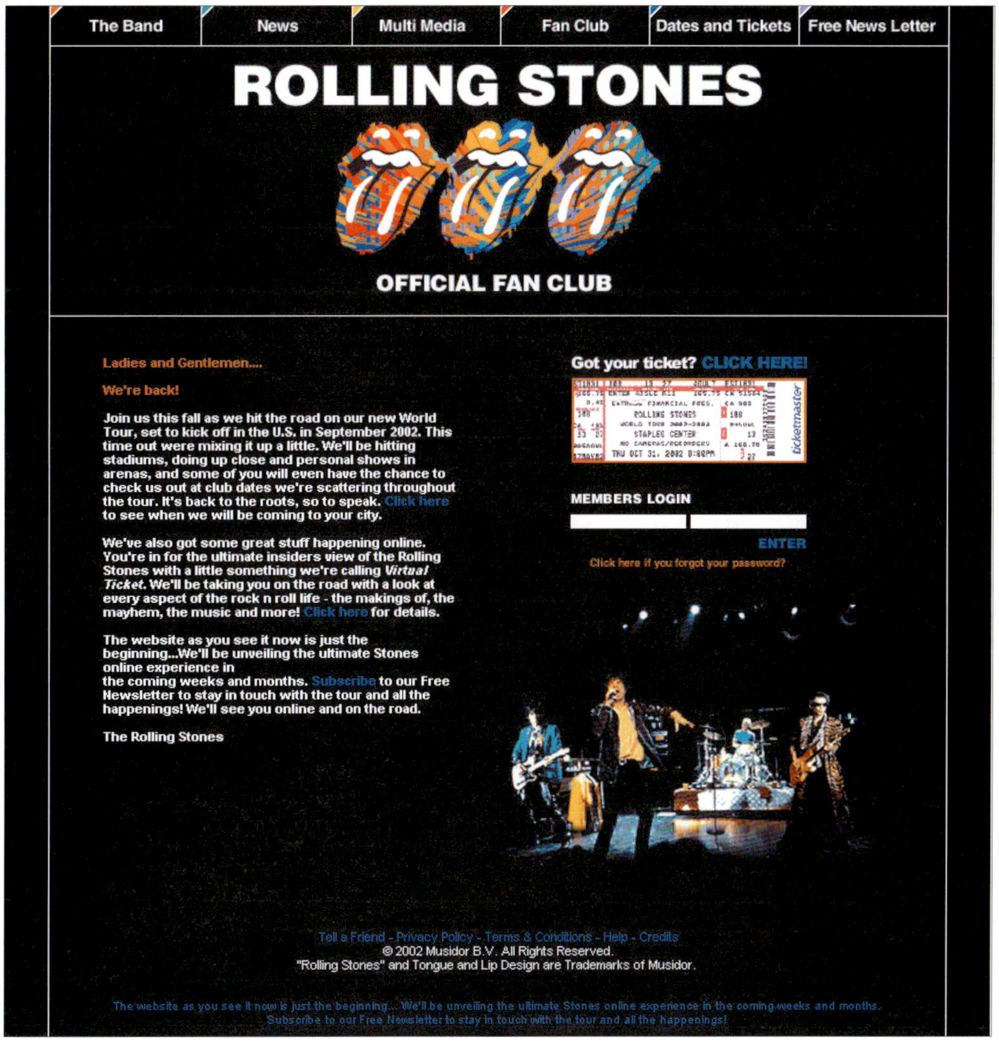

● Rolling Stones
● http://www.rollingstones.com/
● USA／2002
CL, SB：Rolling Stones

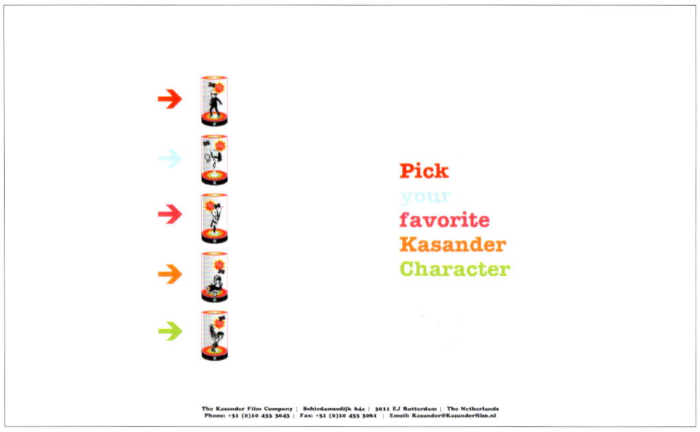

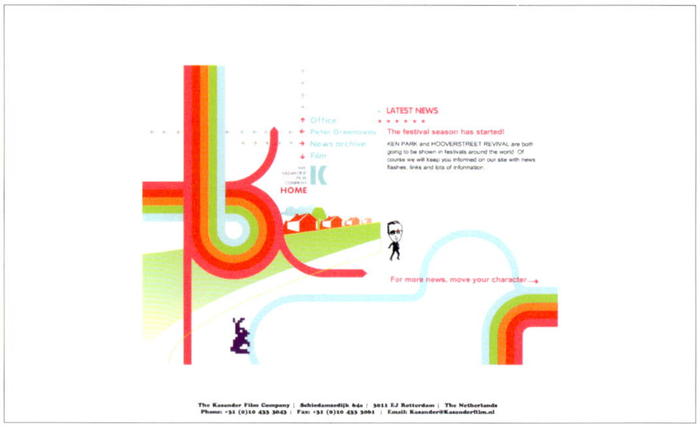

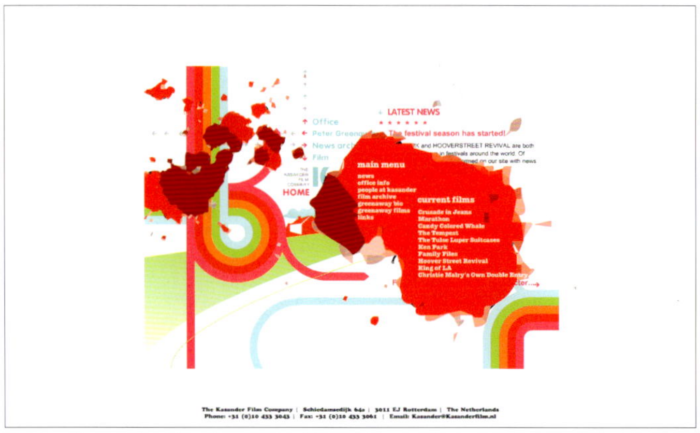

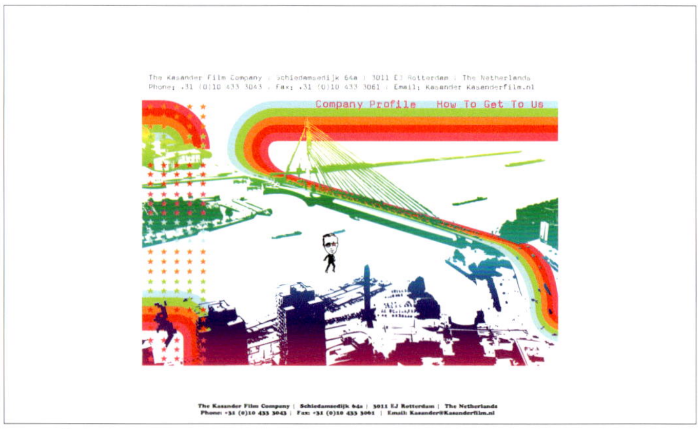

● **The Kasander Film Company Site** ● **http://www.KasanderFilm.nl** ● THE NETHERLANDS／2002

CL：The Kasander Film Company SB：1127.COM DF：Martijn van Egmond ／ Gijs Determeijer of 1127.COM

concept 型破りで有名な映画プロデューサーKees Kasanderは、月並みな映画紹介サイトでは満足しません。70年代に流行したピンボールマシンと香港の広告をモチーフにスタイルを構成し、「不思議の国のアリス」のウサギ、旧式のビデオゲーム、かの有名なHi-Res!の作品であるWebサイト「Jump Tomorrow」にヒントを得て方向キーで移動するナビゲーションをつくりました。登場するキャラクターはすべて、Kasanderの従業員がモデルです。訪れる人は、一風変わったナビゲーションにつられ、ついあちこちを探索してしまうことでしょう。さあ、ウサギについていってみましょうか。

Kees Kasander is an unconventional film producer who didn't want a conventional film site. The style was inspired by 70s pinball machines and Hong Kong advertising, and the arrow key navigation by Alice in Wonderland's rabbit, traditional video games, and the Jump Tomorrow site by Hi-Res!. The characters are the people that work at Kasander. The navigation forces people to wander around and explore; the rabbit is the easy way out...

● SUBAKT ● http://www.subakt.fr
● FRANCE／2001
CL, SB：SUBAKT DF：THE SUBAKT TEAM

concept イラストを中心とした、やわらかなトーンの上品なWebサイト。雑誌のように構成されています。控えめなメニューとは裏腹に、いったんクリックするとプロジェクトが迫力ある全画面で表示され、訪問者はその世界に没頭することができます。「Laboratory」と「Production」のページは根本的に性質が異なり、「Laboratory」にはSUBAKTの実験的な作品が、「Production」には顧客向けに作成したWebサイト、印刷物、ビデオ製品が展示されています。メニューはシンプルなリスト形式で表示されるため、サブメニューがなくても必要な情報に直接アクセスすることができます。

Subdued and elegant, this new Web site makes the illustrations the focus. The site is characterized by a format that looks and feels like a magazine. Apart from the unobtrusive menu, nothing interferes with the presentation of projects. The site provides a total immersion experience; it steps aside and lets content emerge. The Laboratory section of experimental works created by SUBAKT, and the Production section of web, print, and video productions for clients are separated. The simple navigation system is constructed in a list format that enables direct access to information and dispenses with sub menus.

● デザインの現場(Digi-gen)
● http://www.so-net.ne.jp/dezagen/
● Japan／1998
CL, SB：（株）美術出版社 デザインの現場編集部 Bijutsu Shuppan-Sha, Ltd. Designers' Workshop Editorial Department
DF：セキユリヲ Yurio Seki

concept ユーザーが欲しい情報をすぐ得られるよう、トップページの項目を必要最小限におさえ、動画やサウンドを使わないシンプルな構成に徹しました。

The minimalist menu lets users get to the information they want without delay. The layout is simple and doesn't use animation or sound.

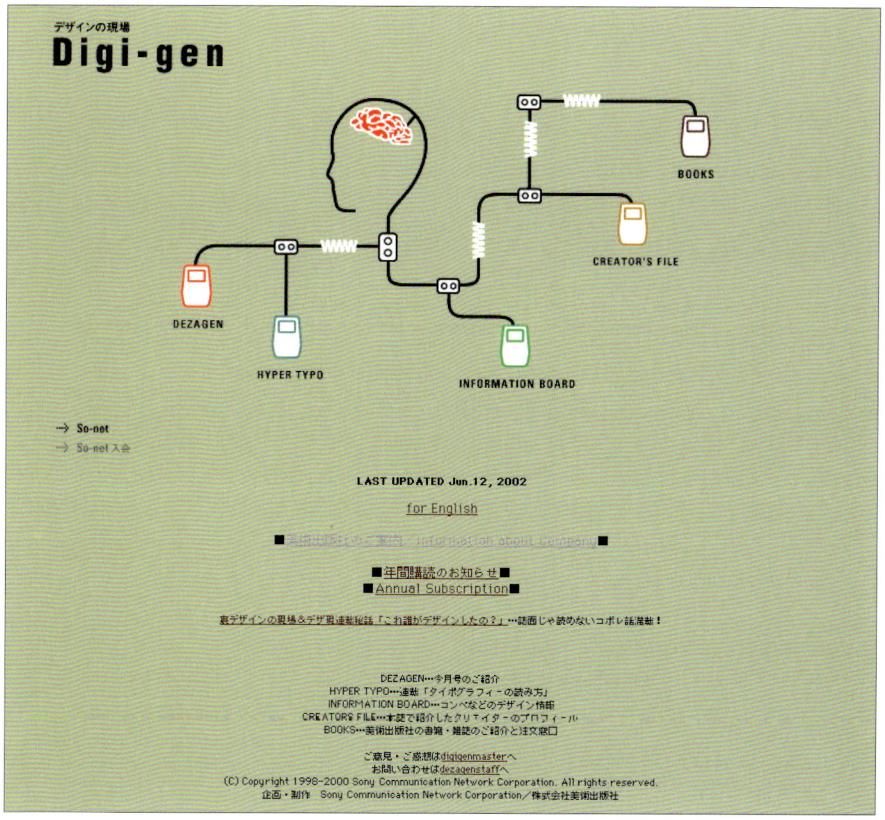

● Petit ● http://www.petit.org/ ● JAPAN／2002

CL, SB：(株)プチグラパブリッシング Petit Grand Publishing, Inc.

DF：marble 梶原綾子 marble Ayako Kajiwara ／ picnica 佐藤岳彦 picnica Takehiko Sato

concept 定期的に更新される情報系コンテンツとオンラインショップ、2つの大きな役割を持ったサイトをいかにシンプルにわかりやすくまとめるかを第一に考えました。特にオンラインショップでは、2002年6月から決済システムを導入。その他にも、受注作業の効率化のために専用アプリケーションを利用するなど、表側だけでなく管理の面でもシステム化を図っています。

This site consists of regularly updated information and an online shop. A special effort was made to design a simple and easy-to-understand site that would encompass these two functions in a compatible fashion. The introduction in June 2002 of a payments system and a dedicated application for ordering products has significantly enhanced the site's effectiveness, not just visually but in terms of improved site management as well.

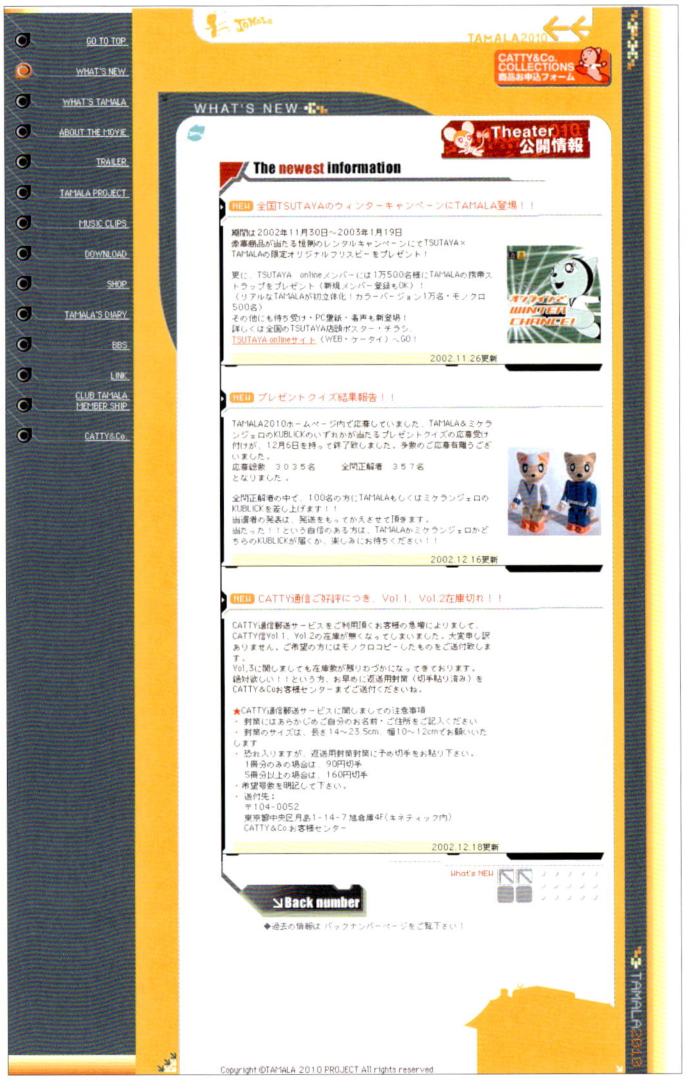

● TAMALA2010 ● http://www.tamala2010.com/tamala/02-whats.html

● JAPAN／2002

CL, SB：(株)キネティック KINETIQUE Inc. DF：キッズコーポレーション(株) KIDS Corporation

● **die gestalten** ● **http://www.die-gestalten.de**
● **GERMANY／2002**
CL, SB： Die Gestalten Verlag GmbH　DF： It's immaterial , Berlin / Die Gestalten Verlag

concept リニューアルオープンしたDie GestaltenのWebサイト。デザイナーとクライアントにとっての市場であり、販売店とメディア関係者にとって使いやすい情報ツールを目指して作成しました。また、ノート型コンピュータの画面でも見やすい構成に仕上げています。

We've relaunched the Die Gestalten Web site. The intention was to create both a marketplace for designers and clients and an informative and easy-to-use tool for distributors and the press. We designed the pages to work well on a laptop screen.

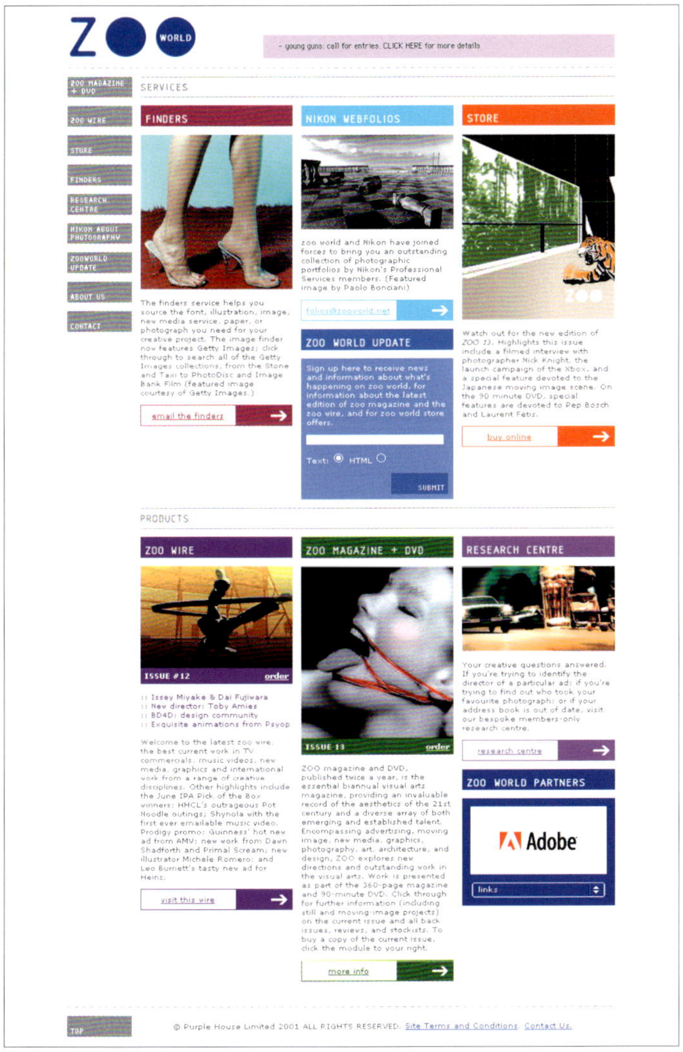

● **ZOO WORLD** ● **http://www.zooworld.net/** ● **2002**
CL, SB： Purple House Limited

● **ARCHINECT** ● **http://www.archinect.com** ● **CANADA／1997**
CL, DF： Paul Petrunia　SB： ARCHINECT

concept 独自のインターネット環境を活用し、世界中のデザイナーから提案された新しいアイデアを紹介する場を設けました。

The ARCHINECT site takes advantage of the unique Internet environment to serve as a venue where designers all over the world and from all disciplines can share ideas in order to make architecture more connected and open minded.

● Style ● http://style.jp/ ● JAPAN／2001
CL, SB：（株）講談社「Style」編集部 KODANSHA Ltd.「Style」Editorial department

● Grazi@ ● http://www.joseishi.net/grazia/index.html ● JAPAN／2002
CL, SB：（株）講談社 Grazia［グラツィア］ KODANSHA Ltd. Publishers Grazia
DF：（株）ラナデザインアソシエイツ RaNa design associates, inc.

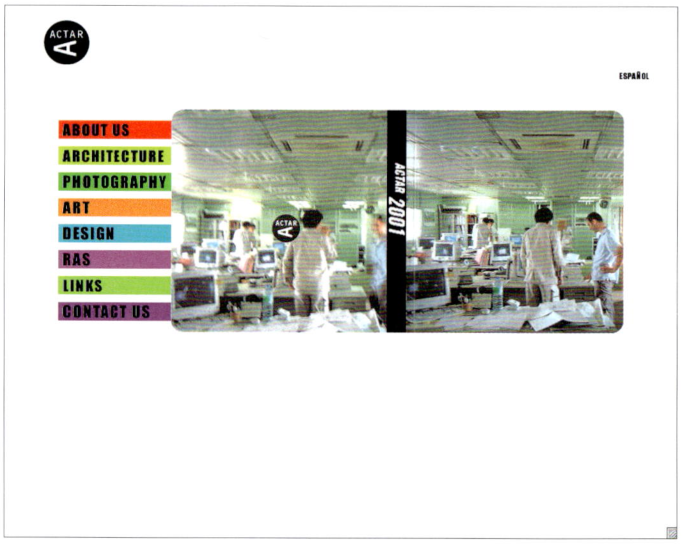

● ACTAR ● http://www.actar.es ● SPAIN／2000
CL SB：Actar DF：坂本知子 Tomoko Sakamoto

concept アクタールが毎年発行している小さなカタログのフォーマットをそのまま踏襲し、ポストイットに見立てたタグで各ページを紹介しています。

The Web site format exactly imitates the catalog Actar issues each year. Users browse through the little catalog by clicking on Post-it index tabs.

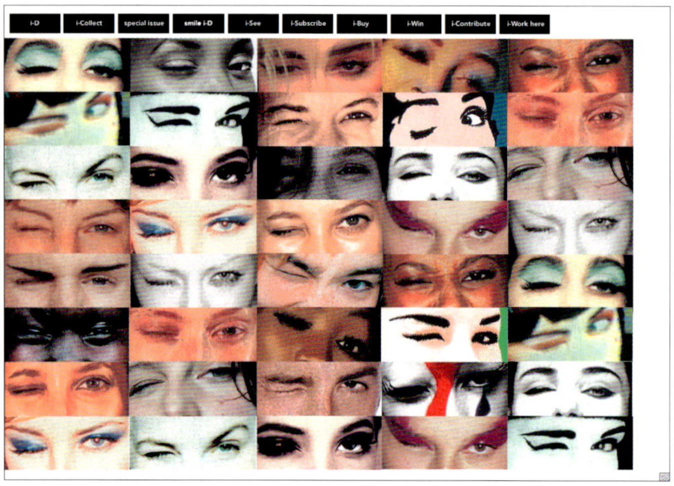

● i-D Magazine ● http://www.i-dmagazine.com ● UK／2001
CL：i-D Magazine. Levelprint ltd SB：i-D DF：Matthew Hawker

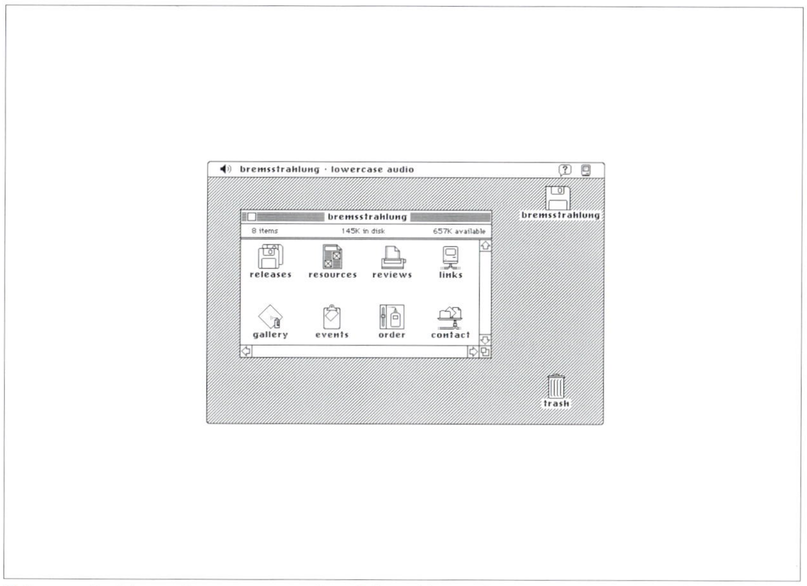

● bremsstrahlung・lowercase audio ● http://www.lowercasesound.com/finder.html
● USA／2002
CL, SB： bremsstrahlung

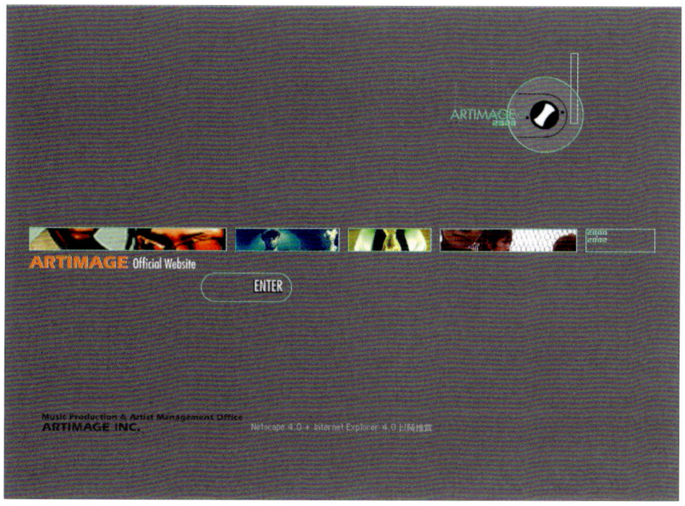

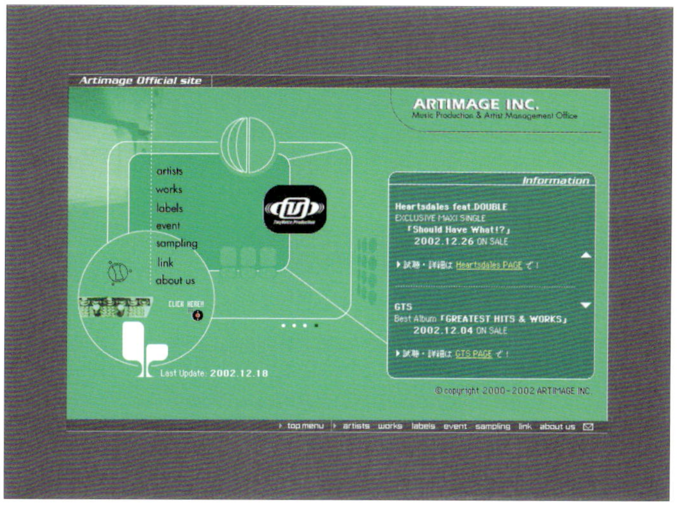

● ARTIMAGE ● http://www.artimage.co.jp ● JAPAN／2000
CL, SB：（株）アーティマージュ ARTIMAGE INC. DF：（株）キウイ KIWI INC.

● Songline/tone field productions
● http://www.songtone.com/default.htm ● USA／2002
CL, SB： Songline / Tone Field Productions

● WHOOP ● http://www.whoop.co.jp ● JAPAN／2002
CL, SB：（株）フープ WHOOP

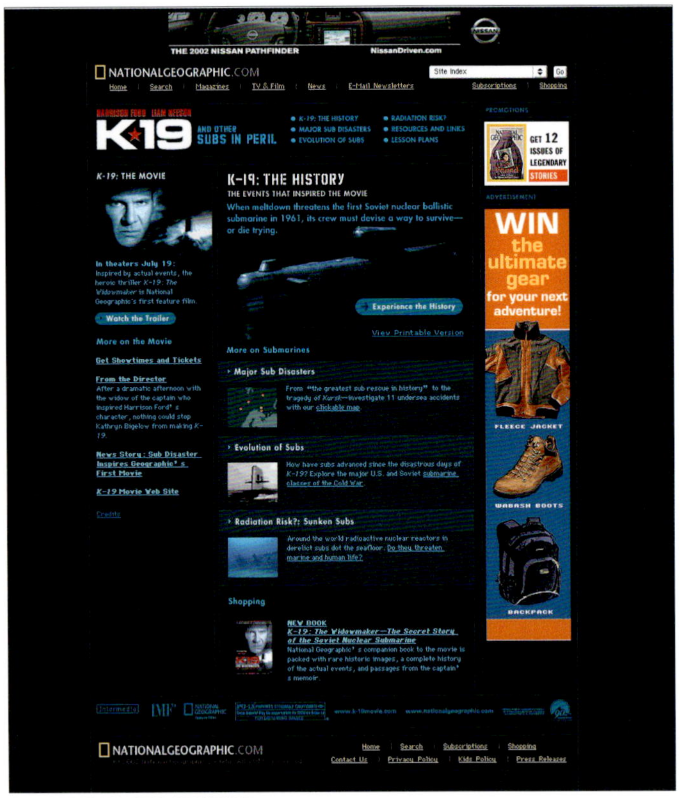

● NATIONALGEOGRAPHIC
● http://crater.nationalgeographic.com/k19/ ● USA／2002
CL, SB：National Geographic Society

● SEIKODO MUSIC GALAXY ● http://www.seikodo.co.jp ● JAPAN／2002
CL, SB：(株)星光堂 SEIKODO Co., Ltd.
DF：東京書籍印刷(株) Tokyo Shoseki Printing Co., Ltd.

● ワーナー・ホーム・ビデオ(WARNER HOME VIDEO) ● http://www.whv.jp
● JAPAN／2002
CL, SB：ワーナー・ホーム・ビデオ Warner Home Video DF：石川潤治(ディレクション) Junji Ishikawa (Director)／宮脇正臣(ディレクション) Masaomi Miyawaki (Director)／加藤優子(デザイナー) Yuko Kato (Designer)

concept 自社タイトルの情報をどこよりも速く、正確に、そして楽しく提供していくことを主眼として更新しております。強力なデータベースを軸として、読み物やプレゼントなど閲覧者にとって益のあるサイトになっております。今後は弊社の作品紹介にとどまらず、ユーザーも参加できる映画・DVDのコミュニティを構築していきます。

Our main consideration in updating the site is to provide users with more accurate, more entertaining information on titles in the Warner Home Video catalog, and get it to them faster than anywhere else. We've tried to create a site visitors will find helpful and beneficial to browse by including a powerful database, reading material, present offers, and more. We are in the process of expanding and developing the site into an interactive movie and DVD community.

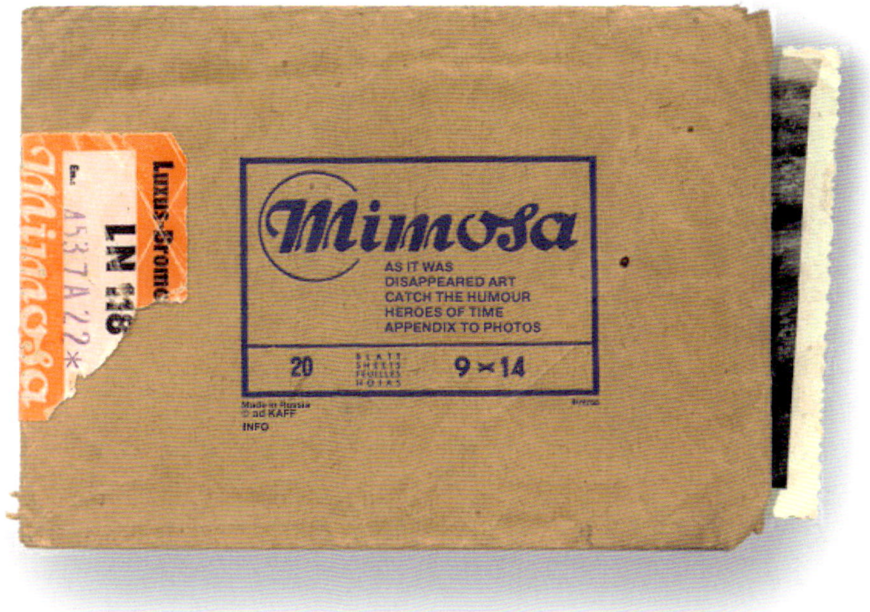

● Mimosa ● http://www.kaff.ru/mimosa/ ● 2002
CL, SB： ad KAFF DF： ad KAFF, kaff.ru

concept ロシアの古い時代の写真ギャラリーです。
The site presents old photos from Russia.

● Musée des arts et métiers ● http://www.arts-et-metiers.net/
● FRANCE／2002
CL, SB： Musée des arts et métiers-CNAM

● ギャラリーなつか（GALLERY NATSUKA）
● http://www.ginza.co.jp/natsuka/ ● JAPAN／2002
CL： ギャラリーなつか GALLERY NATSUKA Co.,ltd. SB, DF： （株）インターカタログネット inter-Catalog-net Inc.

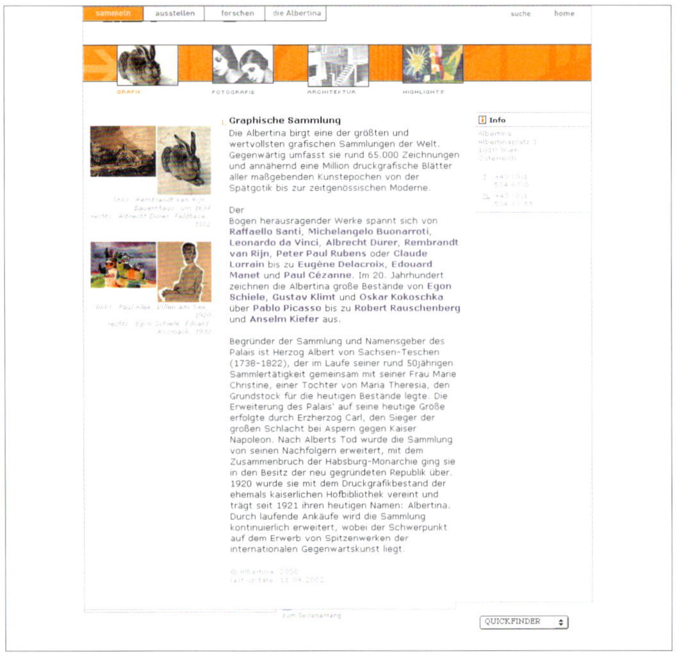

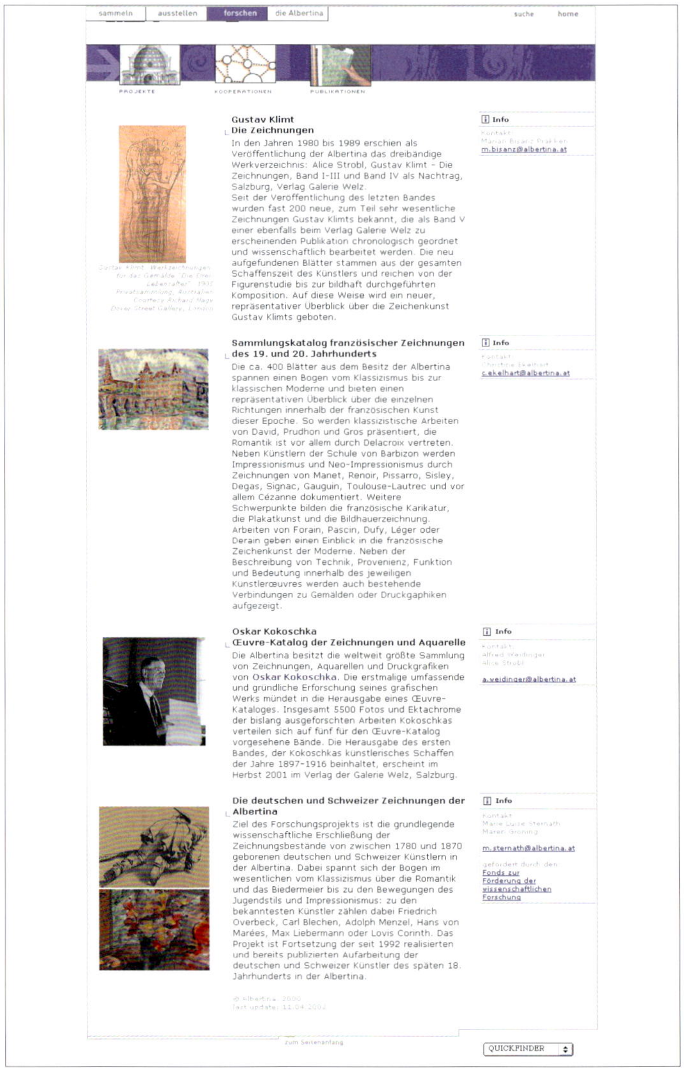

● ALBERTINA ● http://www.albertina.at/index_d_fl.html
● AUSTRIA／2002
CL, SB：Albertina

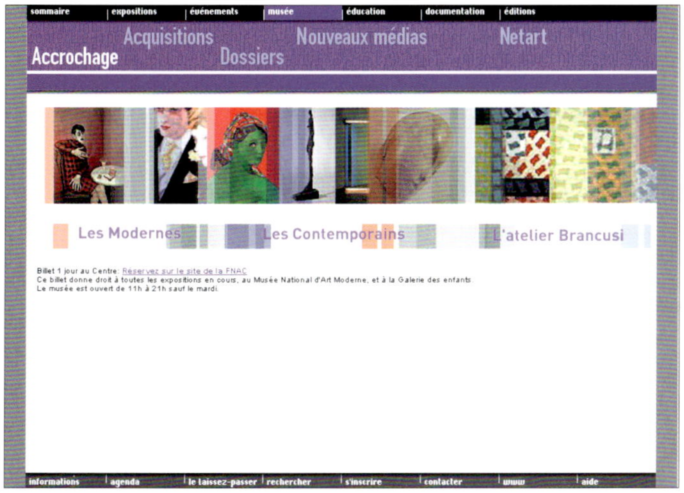

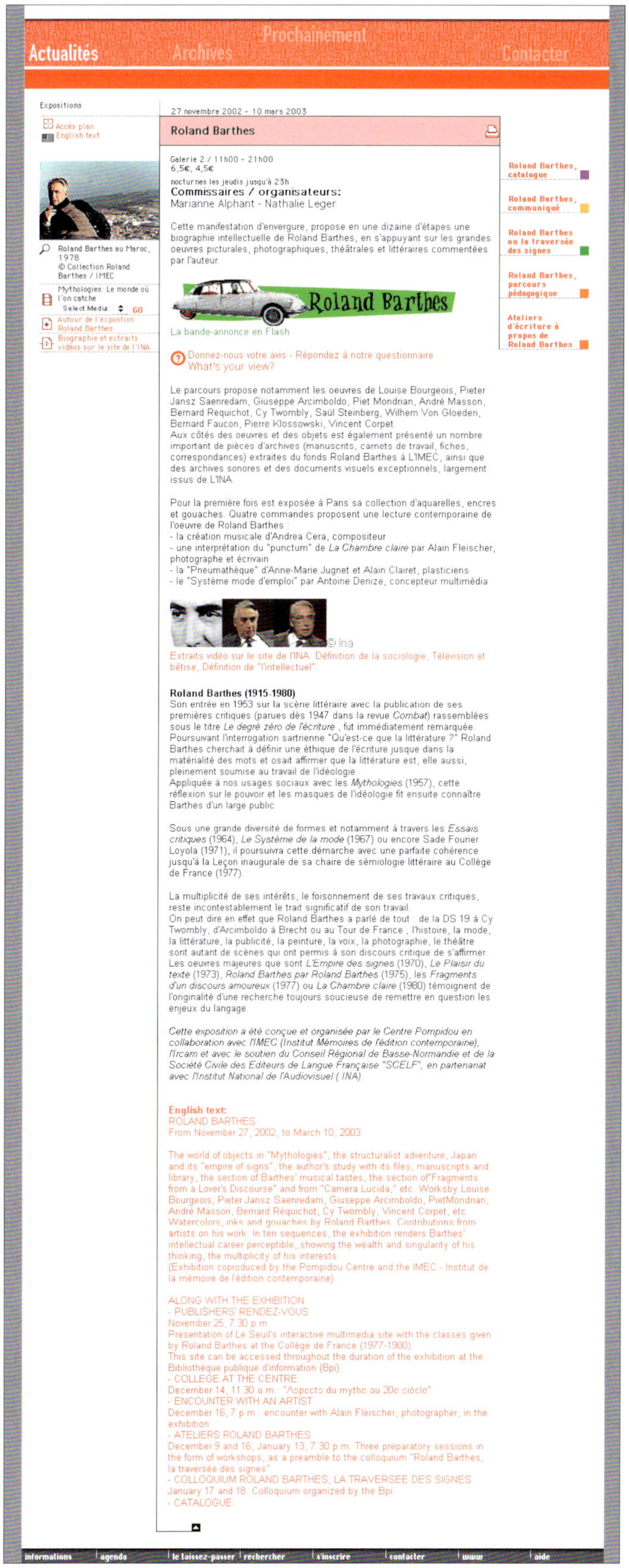

● **Centre Pompidou** ● **http://www.cnac-gp.fr/**
● **FRANCE／2002**
CL, SB：Centre Pompidou

● **Bauhaus** ● **http://www.bauhaus.de**
● **GERMANY／2002**
CL, SB：Bauhaus-Archiv ／ Mueum für Gestaltung
DF：veronika bongarz

concept バウハウス（1919年開校、1933年閉校）のデザイン、建築、芸術分野の遺産を展示するWebサイト。展示されているコレクションは予備課程で制作されたもので、家具、メタルオブジェクト、広告、織物、陶磁器、写真、建築物、絵画にまで多岐にわたります。第一次大戦後、デザイン、建築および芸術の教育機関として設立されたバウハウスのプログラムと作品は、今の時代においてもデザイン開発に多大な影響を与え続けています。バウハウスの革新的な理念と概念の相互作用が生んだ創造力相互作用により、バウハウスは1920年代のデザイン分野においてきわめて重要な存在となったのです。

The Bauhaus Archive / Museum of Design is dedicated to presenting and researching the history and impact of the Bauhaus (1919-33) on design, architecture and art. The extensive collection includes furniture, metal objects, advertisements, weaving, ceramics, photography, architecture, painting, graphic works, and studies from the school's preparatory courses. The Bauhaus was the most important school for design, architecture, and art of the 20th century. Its program and products continue to influence the development of contemporary design. The productive cross-fertilization of innovative ideas and concepts made the Bauhaus a vital center for design in the 1920's.

bauhaus-archiv museum für gestaltung
bauhaus-archiv gebäude museum sammlungen bauhaus 1919-33 aktuelles information bauhaus-shop
geschichte
ansichten
pläne
Gebäude
Das Museumsgebäude ist ein Spätwerk des Bauhaus-Gründers Walter Gropius. Geplant seit 1964 für einen Standort in Darmstadt, wurde das Gebäude 1976-79 in modifizierter Form in Berlin errichtet. Seine markante Silhouette ist heute eines der Wahrzeichen der Stadt
Bauhaus-Archiv/Museum für Gestaltung
impressum

bauhaus-archiv museum für gestaltung
bauhaus-archiv gebäude museum sammlungen bauhaus 1919-33 aktuelles information bauhaus-shop
bauhaus-ausstellung
sonder-ausstellungen
ausstellungs-archiv
Museum
Bauhaus-Archiv / Museum für Gestaltung
Klingelhöferstraße 14
D - 10785 Berlin (Tiergarten)
Das Museum ist täglich, außer dienstags, von 10.00 - 17.00 Uhr geöffnet.
Das Bauhaus-Archiv zeigt in einer ständigen Bauhaus-Ausstellung Ausschnitte seiner Sammlung. Wechselnde Sonderausstellungen präsentieren einzelne Künstler oder aktuelle Themenbereiche aus Kunst, Architektur und Design der Moderne.
VERMERK: Erhalten Sie Informationen zum Bestand der Bauhaus-Archiv-Bibliothek über den Bibliothekskatalog.
Bauhaus-Archiv/Museum für Gestaltung
impressum

bauhaus-archiv museum für gestaltung
bauhaus-archiv gebäude museum sammlungen bauhaus 1919-33 aktuelles information bauhaus-shop
vorgeschichte
manifest
unterricht
werkstätten
architektur
kunst
bühne
fotografie
nachfolge
zeittafel
biographien
Bauhaus 1919-33
Das Bauhaus begann 1919 mit der Formulierung einer Utopie: Der "Bau der Zukunft" sollte alle Künste in idealer Einheit verbinden. Dies erforderte einen neuen Typ des Künstlers jenseits akademischer Spezialisierung, dessen Erziehung am Bauhaus angestrebt wurde. Sein Gründer Walter Gropius sah den Weg zu diesem Ziel in neuen pädagogischen Methoden und im Handwerk als Voraussetzung jeder Kunst: "Die Schule soll allmählich in der Werkstatt aufgehen". Folglich arbeiteten am Weimarer Bauhaus Künstler und Handwerker gemeinsam in Lehre und Produktion. Auf diese Weise sollte die Trennung zwischen freier und angewandter Kunst aufgehoben werden.
Die Realität der technischen Zivilisation stellte jedoch weitergehende Ansprüche, denen die Aufwertung des Handwerks allein nicht genügen konnte. Das Bauhaus reagierte 1923 mit einem geänderten, seine künftige Arbeit bestimmenden Programm unter dem Leitsatz: "Kunst und Technik - eine neue Einheit". Die Möglichkeiten der Industrie sollten für das Ziel einer funktional und ästhetisch befriedigenden Gestaltung eingesetzt werden. In den Bauhauswerkstätten entstanden Vorbilder, die für die Massenproduktion bestimmt waren: von der Lampe bis zum Wohnhaus.
Der erzieherische und soziale Anspruch einer neuen Gestaltung des Lebens und der Umwelt, der hiermit verbunden war, konnte sicherlich nicht immer eingelöst werden. Auch stand das Bauhaus mit dieser Zielsetzung nicht allein. Aber es wurde zum Leitbild, fast zum Synonym solcher Bestrebungen. Die Geschichte des Bauhauses verlief nicht geradlinig. Wechsel in seiner Leitung und unter den Lehrern, künstlerische Einflüsse von außen und die politische Situation, in der das Experiment Bauhaus stattfand, bewirkten seinen ständigen Wandel. Die vielfältigen Wirkungen dieses Experiments reichen bis in unsere Gegenwart.
impressum

bauhaus-archiv museum für gestaltung
bauhaus-archiv gebäude museum sammlungen bauhaus 1919-33 aktuelles information bauhaus-shop
ausstellungen
veranstaltungen
pressetexte
neuerwerbungen
öbel - Eine Legende wird besichtigt +++ 31. Oktober 2002 bis 10. März 2003 -
VERMERK: Informationen zum Bestand der Bauhaus-Archiv-Bibliothek erhalten Sie über den Bibliothekskatalog.
Breuer-Urteil rechtskräftig
Das Urteil des Oberlandesgerichts Düsseldorf vom 30. April 2002 zu Marcel Breuers 1925 entworfenem Stahlrohrhocker ist jetzt rechtskräftig geworden. Wegen seiner grundsätzlichen Bedeutung hatte der Prozeß TECTA ./. L + C Stendal in Presse und Öffentlichkeit erhebliche Aufmerksamkeit gefunden. Nunmehr wurde eine Nichtzulassungsbeschwerde beim Bundesgerichtshof zurückgenommen.
Stahlrohrhocker, 1925
Bauhaus-Archiv, Berlin am 22. Oktober 2002
Das Oberlandesgericht Düsseldorf hat am 30. April 2002 sein Urteil in Sachen TECTA ./. L + C Stendal gesprochen. In dem in der Fachöffentlichkeit mit Spannung verfolgten Prozeß ging es um Marcel Breuers 1925 entworfenen Stahlrohrhocker, dessen urheberrechtliche Schutzfähigkeit sowie um die Frage, wem die Verwertungsrechte zukommen. Das Gericht hat nunmehr der Berufungsklage der Firma TECTA, Lizenznehmerin der Bauhaus-Archiv GmbH, gegen die Firma L + C Stendal stattgegeben. Bei der Urteilsverkündung am 30.04.02 wurden folgende Entscheidungsgründe als maßgebend hervorgehoben:
• Das Gericht hat den Stahlrohrhocker für urheberrechtlich schutzfähig erachtet. Es ist dabei von den Verhältnissen der 20er Jahre ausgegangen, in denen Stahlrohrmöbel dieser Art etwas völlig Neues waren. Dies entspricht auch der Überzeugung des Bauhaus-Archivs. Breuers Stahlrohrhocker kann als ein Werk der angewandten Gestaltung, das auf industrielle Fertigung angelegt ist, sehr wohl künstlerischen Charakter besitzen. Das Werk entspricht in besonderer Weise der von Walter Gropius programmatisch geforderten Verbindung von Kunst und Technik.
• Das Gericht hat ferner festgestellt, daß allein aus dem Anstellungsverhältnis Breuers bei der Stadt Dessau nicht auf eine Rechtsübertragung geschlossen werden könne. Breuer habe den strittigen Hocker nicht im Bereich des Bauhauses geschaffen, auch habe die Entwicklung von Stahlrohrmöbeln nicht zu seinem Aufgabenbereich im Bauhaus gehört. Ebenso wie bei seinen Malerkollegen könne man davon ausgehen, daß nicht alles, was Breuer als Künstler während seiner Bauhaus-Jahre gemacht habe, dem Bauhaus zustehe. Daraus ergibt sich, daß die Nutzung des von ihm persönlich geschaffenen Werks, in diesem Falle des Stahlrohrhockers, Marcel Breuer bzw. seinen Erben zusteht, nicht seinem damaligen Arbeitgeber. Auch diese Feststellungen entsprechen der Überzeugung des Bauhaus-Archivs.
• Schließlich hat das Gericht überprüft, ob die Rechtsübertragung von der Erbin Marcel Breuers auf die Bauhaus-Archiv GmbH und von dieser auf die Firma TECTA wirksam erfolgt sei. Zwar sei die Rechtsübertragung in der ersten Instanz nicht ausreichend nachgewiesen worden, inzwischen ist sie jedoch durch neue Verträge vom 11.04.2001 wirksam erfolgt.
Das Bauhaus-Archiv stellt mit Genugtuung fest, daß seine Rechtsauffassung durch das Urteil des OLG Düsseldorf vom 30. April 2002 in vollem Umfang bestätigt wurde.
Bauhaus-Archiv, Berlin, am 08. Mai 2002
31. Oktober 2002 bis 10. März 2003
"Bauhaus-Möbel"
Eine Legende wird besichtigt
Bauhaus ist zu einer Jahrhundertlegende geworden, die Begriffe ' Bauhaus' und ' modern' werden heute nahezu gleichgesetzt. Dieser Erfolg ist in besonderer Weise den Möbeln zu verdanken, die am Bauhaus oder (was nicht dasselbe ist) von dessen Gestaltern entworfen wurden. Längst werden die Stahlrohrstühle eines Marcel Breuer oder Ludwig Mies van der Rohe als "Klassiker der Moderne" gefeiert. Oft wird jedoch übersehen: das Bauhaus erschöpft sich keineswegs im Material Stahlrohr, und es sind auch nicht ausschließlich Stühle dort entstanden. Unsere Ausstellung wird deshalb ein sehr viel umfassenderes Panorama vorführen: Stühle und Tische, aber auch Betten und Schränke, aus allen Epochen des Bauhauses, aus Holz wie aus Stahl, Welterfolge wie Schülerexperimente. Zugleich weisen wir auch auf die andere Moderne hin, die sich neben dem Bauhaus in Deutschland entfaltete.
Die gezeigten Möbel sollen aus ihrer Epoche, den 20er und frühen 30er Jahren, interpretiert werden. Wir untersuchen, was damals im Umkreis des Bauhauses ent-stand, wer die Möbel in Auftrag gegeben hat und wie sie verwendet wurden. Dazu sind zeitgenössische Zeitschriften und Filme eine unverzichtbare Quelle. Kein "Möbelhaus Bauhaus" wird es geben, sondern ein dif-ferenzierendes "Zeit-Bild".
Bei der Realisierung stützt sich das Bauhaus-Archiv auf seine in über vierzig Jahren aufgebaute Sammlung. Eine Reihe wertvoller Leihgaben wird die Ausstellung ergän-zen. Sie wird von einem Katalog begleitet. Das Projekt wird vom Hauptstadtkultur-fonds und der Kulturstiftung der Länder unterstützt.
Presseinformation zur Ausstellung: Bauhaus-Möbel. Eine Legende wird besichtigt
Sie sind an weiterführenden Informationen interessiert? Im Mitglieder-Forum finden sie ergänzende Beiträge zu unseren Ausstellungen
16. April bis 30. Juni 2003
"bauhausleuchten? KANDEMLICHT!"
Die Zusammenarbeit des Bauhauses mit der Leipziger Firma Kandem.
Seit 1928 entstand aus der Kooperation des Dessauer Bauhauses mit der Leipziger Leuchtenfabrik Körting & Mathiesen (Markenname: Kandem) eine Serie von neuartigen Leuchtenmodellen, in denen das Bauhaus-Programm der praktischen Zusammenarbeit mit der Industrie erstmals erfolgreich realisiert werden konnte. Die Leuchten wurden von Marianne Brandt, Hin Bredendieck, Heinrich Siegfried Bormann und anderen Studierenden des Bauhauses gemeinsam mit den Lichttechnikern von Kandem für die Serienproduktion entwickelt und fanden sofort ihren Weg in Wohnungen und Büros. Innerhalb weniger Jahre wurden damals über 50000 aus dieser Partnerschaft hervorgegangene Decken- und Tischleuchten verkauft.
Die vom Museum für Kunsthandwerk / Grassimuseum Leipzig gemeinsam mit dem Gastkurator Justus A. Binroth und unter Beteiligung des Bauhaus-Archivs erarbeitete Ausstellung ist vom 14.11.2002 bis zum 23.2.2003 in Leipzig zu sehen (www.grassimuseum.de) und wird anschließend im Bauhaus-Archiv Berlin gezeigt. Sie präsentiert alle am Bauhaus entwickelten Kandem-Leuchten zusammen mit Vorgängermodellen und Varianten; hinzu kommen Prototypen und Entwürfe von Leuchten, die nie zur Serienreife gelangten. Zeitgenössische Fotografien, Kataloge, Prospekte und andere Werbematerialien dokumentieren das breit gefächerte Produktionsspektrum der Firma Körting & Mathiesen, die damals zu den führenden Herstellern der deutschen Leuchtenindustrie zählte. Mit ihren wissenschaftlichen Entwicklungsmethoden und rationalisierten Produktionstechniken kam sie den gestalterischen Absichten des Bauhauses auf ideale Weise entgegen. Diese für beide Seiten fruchtbare Kooperation, ein wichtiges Kapitel in der Geschichte der modernen industriellen Gestaltung, wird hier zum ersten Mal umfassend dargestellt.
Das Katalogbuch zur Ausstellung (deutsch/englisch) mit Beiträgen von Justus A. Binroth, Katrin Heise, Peter M. Kleine, Ulrich Krüger, Klaus Struve, Olaf Thormann und Klaus Weber ist erschienen bei Arnoldsche Art Publishers, Stuttgart. Preis in der Ausstellung € 24.80
Änderungen vorbehalten / Stand 29.11.2002
impressum

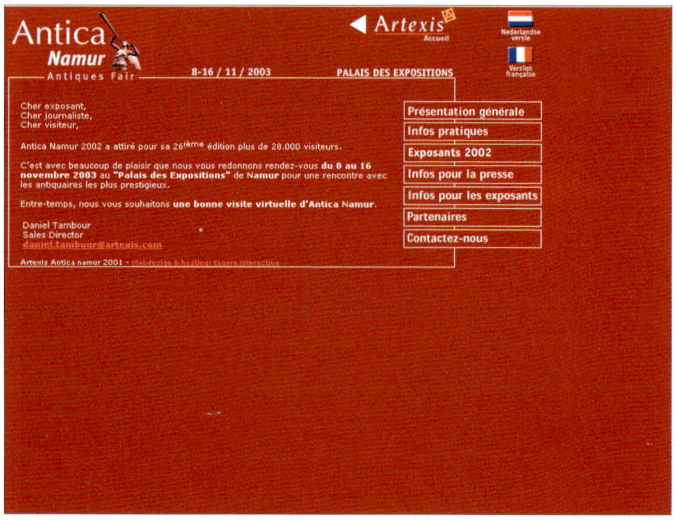

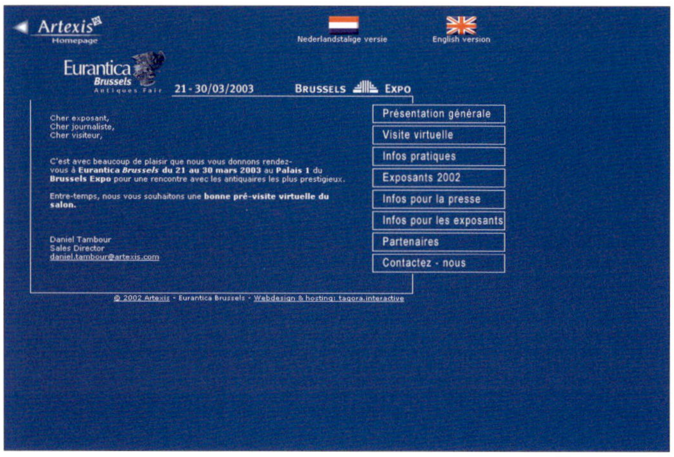

● Artexis ● http://www.artexis.com/home.html ● 2002
CL, SB : Artexis

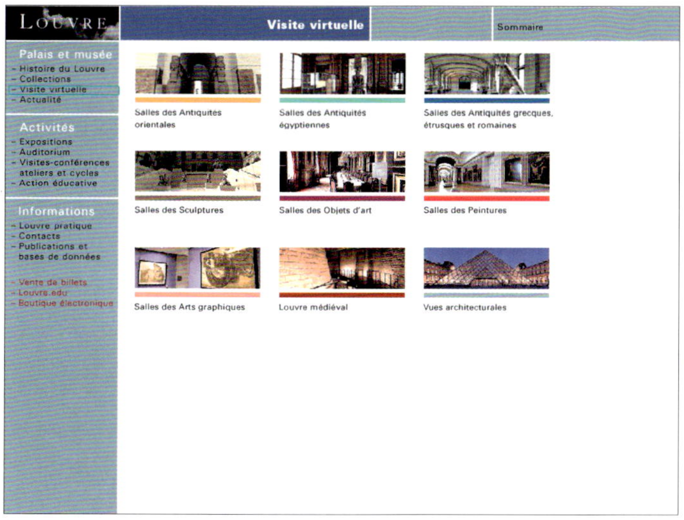

● **LOUVRE** ● **http://www.louvre.fr/** ● **FRANCE／2002**

CL, SB：Louvre Museum

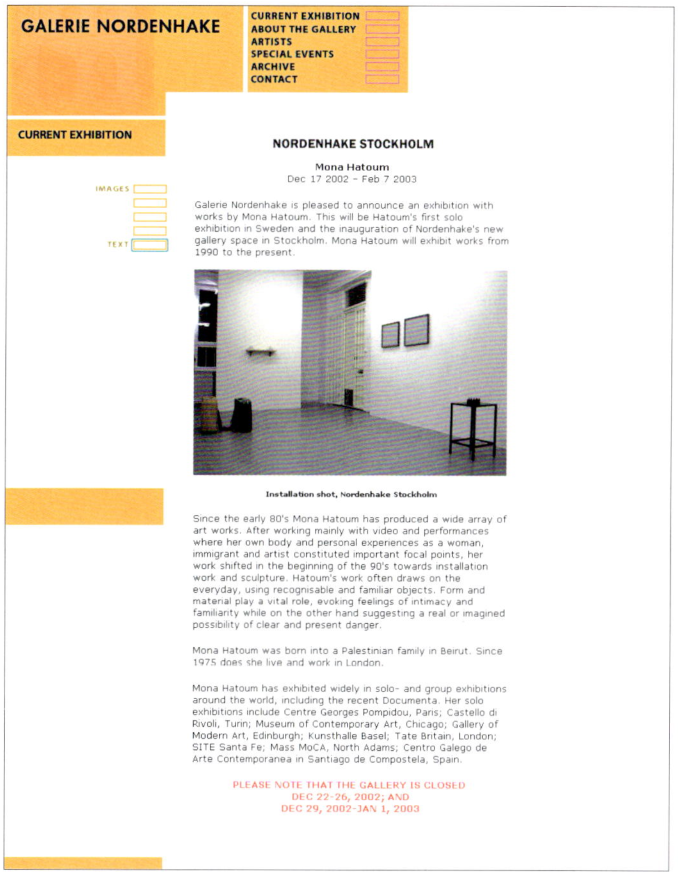

● **GALERIE NORDENHAKE** ● **http://www.nordenhake.com/**
● **GERMANY／2002**

CL, SB：Galerie Nordenhake

OTHERS

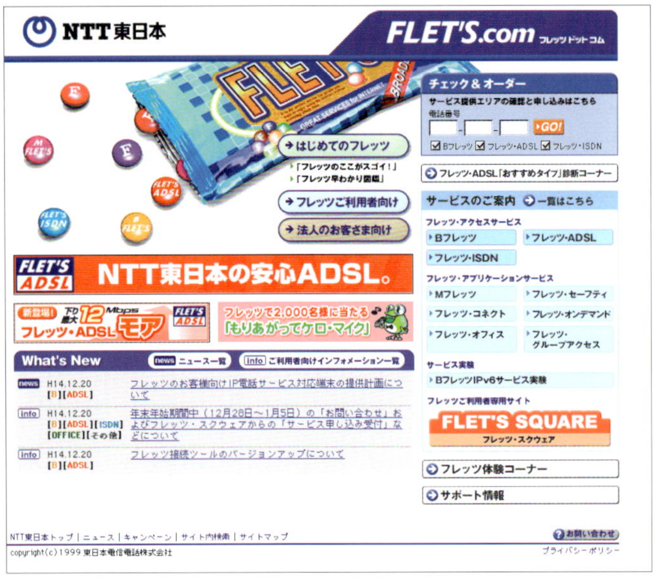

● フレッツ(FLET'S) ● http://flets.com/ ● JAPAN／2002
CL, SB：NTT東日本 NIPPON TELEGRAPH AND TELEPHONE EAST CORPORATION

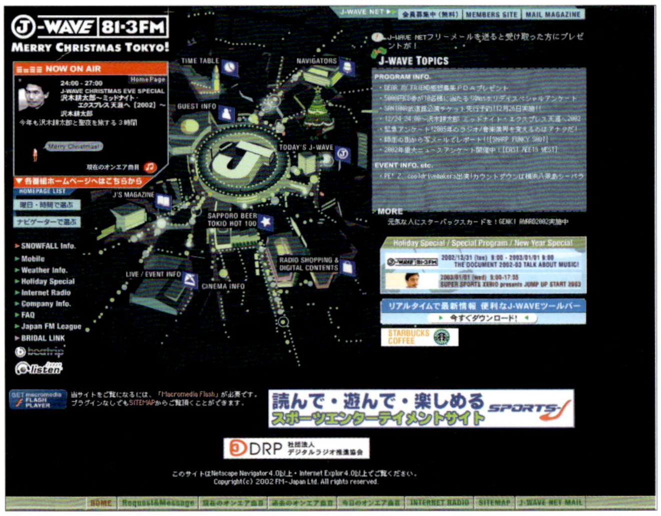

● J-WAVE ● http://www.j-wave.co.jp ● JAPAN／2002
CL, SB：J-WAVE DF：レインボージャパン RAINBOW JAPAN

● 杉野學園（Sugino Gakuen） ● http://www.sugino.ac.jp
● JAPAN／2001
CL, SB：学校法人 杉野学園 Educational Institution Sugino Gakuen
DF：（株）インファス＆NTTネットワーク INFAS & NTT Network Inc. ©杉野学園

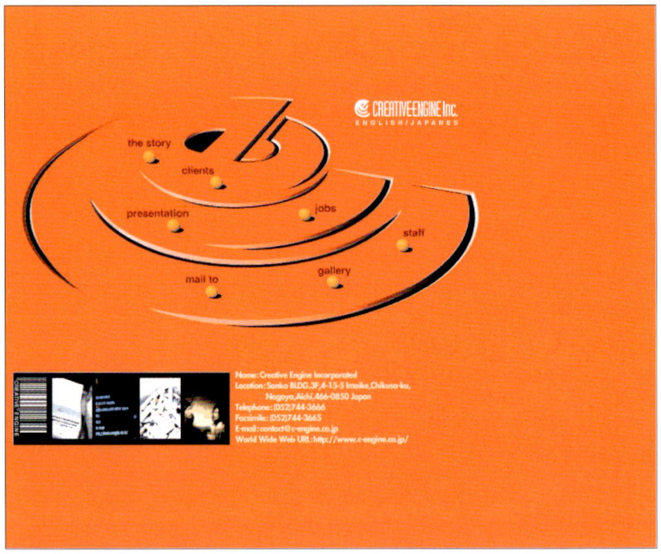

● CREATIVEENGINE Inc. ● http://www.aitcoms.co.jp ● JAPAN／2002
CL, SB：（株）オルタナティブ コミュニケーションズ Alternative Communications Inc.

● Vantan ● http://www.vantan.co.jp/ ● JAPAN／2002
CL, SB, DF：（株）バンタン Vantan Co., Ltd.

● WAVE ● http://www.mswave.co.jp ● JAPAN／2002
CL, SB：（株）マルチメディアスクール・ウェーヴ Multimedia School WAVE
DF：川口愉快 Yukai Kawaguchi

● 森ビル(MORI BUILDING) ● http://www.mori.co.jp ● JAPAN／2001-2002
CL, SB：森ビル(株) MORI BUILDING CO., LTD.

●ホテル西洋 銀座(Hotel Seiyo Ginza) ● http://www.seiyo-ginza.co.jp
● JAPAN／2001
CL, SB：(株)エイチ・エス・ジー ホテル西洋 銀座 H.S.G CO., LTD. Hotel Seiyo Ginza

● クイーンズスクエア横浜(Queen's Square YOKOHAMA)
● http://www.qsy.co.jp ● JAPAN／2002
CL：クイーンズスクエア横浜 Queen's Square Yokohama SB：横浜シティ・マネジメント(株) YOKOHAMA City Managemant Co.,Ltd.

● MID-TOKYO MAPS ● http://www.mid-tokyo.com
● JAPAN／2001-2002
CL, SB：森ビル(株) MORI BUILDING CO., LTD.

● Paletta ● http://www.miyashiro.co.jp/paletta/index.html
● JAPAN／2002
CL, SB：三八城建設(株) Miyashiro Construction, Inc. DF：上鍋紀子 Noriko Uenabe NIWAKO（イラスト） NIWAKO (Illustration)

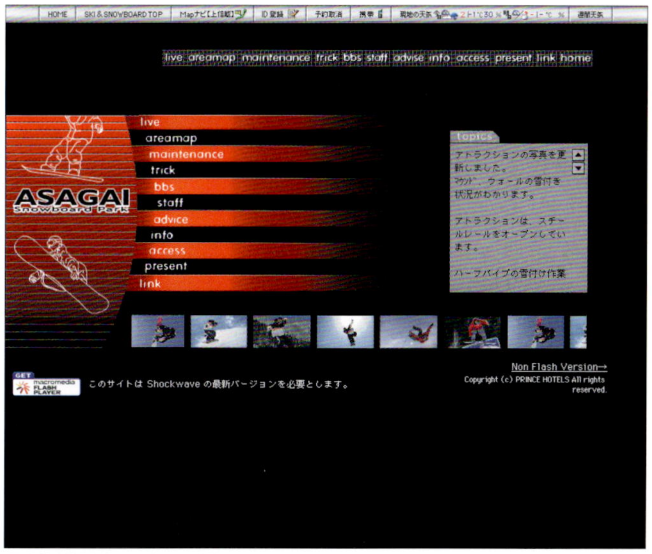

● ASAGAI Snowboard Park
● http://www.princehotels.co.jp/snowboard/asagai/ ● JAPAN／2002
CL SB：(株) コクド 苗場スキー場 Kokudo Co., Ltd. Mt.Naeba SnowResort DF：西武建設(株) 企画開発事業部 企画部 SEIBU construction co.,ltd. Plan development operation division Planning department

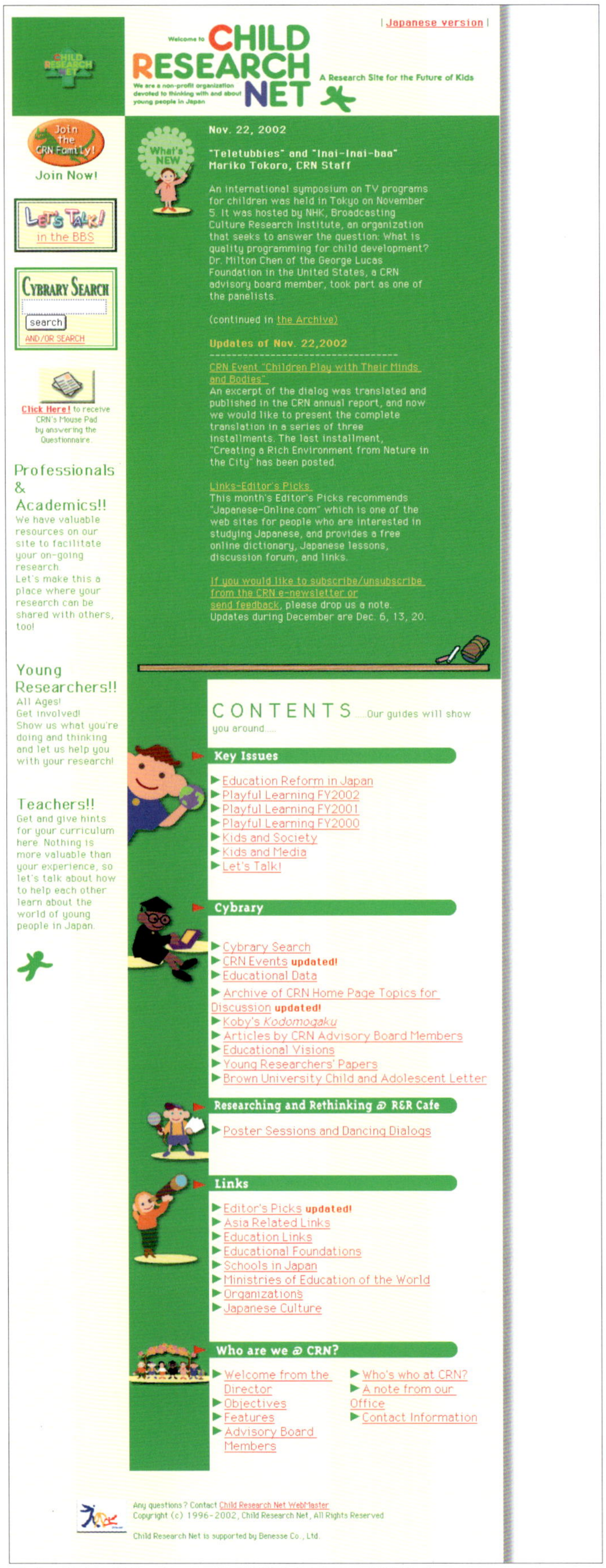

● チャイルド・リサーチ・ネット（CHILD RESEARCH NET）
● http://www.childresearch.net/　● JAPAN／1996
CL, SB：CRN (Child Research Net)

● チャイルド・リサーチ・ネット（CHILD RESEARCH NET）
● http://www.crn.or.jp/　● JAPAN／1996
CL, SB：CRN (Child Research Net)

INDEX

INDEX

INDEX

BY INDUSTRY : WEBSITE MENU COLLECTION
業種別 ウェブメニュー コレクション

Designer
安野　久美子
Kumiko Yasuno

Editor
立古和智
Kazutomo Ryuko

Coodinators
よしひろまさみち
Masamichi Yoshihiro
山本高樹
Takaki Yamamoto
金子真弓
Mayumi Kaneko
斉藤　香
Kaori Saito

Creative Director
世浪　淳
Jun Yonami

Translators
（有）ジェックス
JEX Limited
村田スーザン
Susan Murata

Publisher
三芳伸吾
Shingo Miyoshi

2003年3月15日初版第1刷発行

発行所　ピエ・ブックス
〒170-0003 東京都豊島区駒込4-14-6-301
編集●TEL：03-3949-5010　FAX：03-3949-5650
営業●TEL：03-3940-8302　FAX：03-3576-7361

2003年3月17日から住所・電話番号が変わります。

〒170-0005 東京都豊島区南大塚2-32-4
TEL：03-5395-4811　FAX：03-5395-4812

e-mail●editor@piebooks.com
sales@piebooks.com
URL●http://www.piebooks.com

印刷・製本●（株）サンニチ印刷

ISBN4-89444-235-3 C3070
Printed in Japan